HIST
BRIT N'S RAILWAYS

Published by Collins
An imprint of HarperCollins Publishers
Westerhill Road, Bishopbriggs, Glasgow G64 2QT
www.harpercollins.co.uk

First published as The Times History of Britain's Railways 2015
Paperback Edition 2017

A catalogue record for this book is available from the British Library

978 0 00 822896 5

10 9 8 7 6 5 4 3 2 1

Printed in China by RR Donnelley APS Co Ltd

If you would like to comment on any aspect of this publication,
please contact us at the above address or online.
e-mail: **collinsmaps@harpercollins.co.uk**

facebook.com/collinsmaps

@collinsmaps

HarperCollins Publishers Since 1817

HISTORY OF BRITAIN'S RAILWAYS

JULIAN HOLLAND

Contents

Completed in 1890 and still in regular use today, the Forth railway bridge in all of its glory is a lasting monument to the visionary entrepreneurs and railway builders of the great Victorian age in Britain.

Introduction

Railways have featured in Britain's landscape for more than 400 years. Built in the early seventeenth century, the first primitive wagonways were created to transport coal and other minerals from mines to the nearest rivers using horse-drawn wagons with flangeless wheels running on parallel timber rails. By the late 1700s, these timber lines had been replaced by cast-iron plateways, many of which served as feeders to Britain's expanding network of over 4,000 miles of waterways. This burgeoning transport system offered huge capabilities for moving raw materials and finished goods from one end of the country to the other but was limited by the pace of its horse-drawn power. All this was to change, however, with the invention of the steam engine. One of the most important technological breakthroughs ever witnessed by mankind, the steam engine heralded a railway revolution, sealing our position as the pioneers of modern rail transport and changing the face of our world forever.

Cornishman Richard Trevithick's pioneering high-pressure steam railway locomotive made its début in 1804 when it hauled a loaded train on the Penydarren plateway in South Wales. Steam locomotives designed by John Blenkinsop, William Hedley and Timothy Hackworth for use on colliery railways in northeast England soon followed but these primitive machines were rapidly eclipsed by the work of George Stephenson, a self-taught genius who many consider to be the 'Father of Railways'.

Like his predecessors, Stephenson began by building locomotives for use on colliery railways but his vision led him further, to the building of the Stockton & Darlington Railway, the world's first public railway to use steam locomotives. Opening in 1825, this was followed in 1830 by the first inter-city railway, the Liverpool & Manchester Railway. Stephenson's son, Robert, was to share his passion for steam power, designing locomotives such as the *Rocket*, which went on to win the 1829 Rainhill Trials with a hitherto unprecedented speed of 30 mph. From then on the evolution of the steam locomotive was rapid, culminating in 1938 with a world speed record of 126 mph set by Nigel Gresley's *Mallard*, a record that still holds today.

The success of the Liverpool & Manchester Railway opened the floodgates for new railway proposals across Britain. While many of these, such as the Grand Junction and the London & Birmingham railways, went on to form the nucleus of today's modern railway network, others were ill conceived and even downright fraudulent. A frenzy of speculation by investors large and small saw 'railway mania' reach its peak in 1846 when 272 Acts of Parliament authorized 9,500 miles of new railway, many of which were never built. Countless investors lost their life savings and the unscrupulous villain of the piece, 'Railway King', George Hudson, was ousted from his empire a broken man.

As the nineteenth century progressed, so did Britain's railway network and by the early twentieth century it had reached virtually every corner of the land – at its peak extending to around 23,000 route miles and employing around 600,000 people. Combined with Victorian entrepreneurial spirit, limitless supplies of coal and cheap labour, steam railways were the foundation stones that underpinned Britain's industrial revolution, securing its position, by the early twentieth century, as the most powerful nation in the world.

The first major threat to Britain's rail network came in 1914. As the possibility of war loomed large, the Government brought the railways under their control in a move to protect such a valuable strategic resource. They survived the war but emerged in a rundown state and with industrial unrest bubbling to the surface, the Government was forced to take further measures to protect their future. This they did by grouping the country's 120 railway companies together into just four big monolithic and geographical entities. The 'Big Four' Grouping of 1923 led to creeping modernization such as the Southern Railway's third-rail electrification scheme, the standardization of steam locomotive types and most importantly, an intensification of competition that resulted in the 'Golden Age' of high-speed rail travel. All this was to come to an abrupt end on 1 September 1939 when Britain was, once again, at war with Germany.

Over the next six years Britain's railways sustained enormous damage from enemy bombing and overloading due to transporting vast quantities of war materials, especially in the run-up to D-Day. As in the First World War, women took on many previously male-only roles on the railways as their menfolk fought overseas. Once again, most were summarily dismissed at the war's end and those that remained were underpaid and treated as second-class employees – a situation that would change little for almost another thirty years.

STEPHEN
CANN
&
CO. LTD.

The railways were finally nationalized in 1948 with the creation of British Railways (BR). Modernization was desperately needed but with the country close to bankruptcy in the post-war period, progress was painfully slow. BR had inherited a motley collection of steam locomotives, some of them dating back to the late Victorian era. A total of 999 new, more efficient locomotives were constructed between 1951 and 1960 but a modernization plan announced in 1955 was to spell the end for BR's 19,000 steam locomotives – some of them only a few years old – by replacing them with hurriedly introduced diesel-electric and diesel-hydraulic models.

Around 3,000 miles of loss-making rural railways were closed between 1948 and 1962, and with advances in road transport coming on apace and BR plagued throughout the mid-1950s by damaging industrial disputes, more and more freight traffic was being permanently lost to road haulage. Britain's railways were running at a loss to the taxpayer and this deficit kept growing year on year.

Enter Dr Richard Beeching, then technical director at ICI, who was appointed by a pro-road, anti-rail Conservative government to wield his axe and return the railways to profit. Published in 1963, the notorious 'Beeching Report' resulted in the closure of some 2,500 stations and over

PREVIOUS PAGE: The end of an era captured by railway photographer Colin Gifford – a Stanier 'Black Five' 4-6-0 lifts a westbound freight train across a viaduct in Burnley, Lancashire, on 7 July 1968, just over one month before the eradication of standard-gauge steam on British Rail.

BELOW: The sun is just setting, lighting up the northern sky with its warm colours, somewhat at odds with the temperature on the ground but curiously matching the colours of the Colas Railfreight liveried Class '66' No. 66849 *Wylam Dilly*, breasting the 1,169 feet above sea level summit at Ais Gill. The sky is as black as night beyond to the south and the driver of the Carlisle to Chirk timber train faces hail ahead, battering onto his cab windows – the temperature on 3 December 2012 hovering just slightly above zero.

4,000 miles of railway by the mid-1970s, together with the loss of nearly 68,000 jobs. The eradication of standard-gauge steam haulage on BR was achieved in 1968 and the railways entered a supposedly new, shiny, efficient and modern era. With the likes of Inter-City and an expanding electrification programme in progress, the future for the reborn British Rail started to look more promising.

Widely seen as a prelude to privatization, the six regions of BR were replaced by business sectors in 1982. The Serpell Report of 1983, which suggested closing 80 per cent of Britain's railways, was quietly forgotten by the Conservative government. The Railways Act of 1993 finally brought an end to forty-five years of rail nationalization, ushering in a new era of train-operating companies (TOC), freight-operating companies (FOC) and rolling stock operating companies (ROSCO). The control of the infrastructure including track, signalling, tunnels, bridges and most of the stations was handed over to Railtrack but, in the wake of several horrific railway accidents, the company was declared bankrupt in 2001. Its role was taken over by Network Rail, a 'not-for-profit' government-created company.

With increased Government investment in infrastructure, rising passenger numbers and steadily increasing freight traffic, the future now looks brighter for Britain's railways than it has for eighty years. However, Britain lags miserably behind many European and Asian countries when it comes to dedicated high-speed railways, despite the opening of the Channel Tunnel in 1994 and HS1 in 2007. Even if HS2 is built between London, the Midlands and the North we will never come close to what other countries such as France and Japan have already achieved. It is a sobering thought that although Britain was the birthplace of railways and was once at the spearhead of technical innovation, two world wars, a lack of investment, labour disputes, ill-conceived decision-making and poor management have seen its railways fall far behind those of other advanced nations.

1604	1748	1758	1781	1803	1804	1807	1808	1812
Wollaton Waggonway opens in Nottinghamshire	Horse-drawn Wylam Colliery railway opens	Middleton Railway first railway to be built under an Act of Parliament	George Stephenson born in Wylam, Northumberland	Horse-drawn Surrey Iron Railway opens	Trevithick's *Penydarren* becomes world's first steam locomotive to run on rails	Horse-drawn Oystermouth Railway first to carry passengers	Trevithick's *Catch Me Who Can* locomotive demonstrated in London	Scotland's first railway, the Kilmarnock & Troon, opens

IN THE BEGINNING

1604–1825

- **1813–1814** George Stephenson builds his first steam locomotive, *Blücher*
- **1813–1814** William Hedley's *Puffing Billy* introduced on Wylam Colliery railway
- **1817** Kilmarnock & Troon Railway first in Scotland to operate steam locomotives
- **1819** Horsedrawn Mansfield & Pinxton Railway opens
- **1822** George Stephenson's Hetton Colliery railway opens
- **1822** George Stephenson appointed chief engineer of Stockton & Darlington Railway
- **1825** Built to the standard gauge of 4 ft 8½ in., the Stockton & Darlington Railway opens

Early wagonways and tramroads

The practice of guiding carts and wagons along ruts cut into rock dates back at least 2000 years to when the Ancient Greeks used this basic technology to transport stone from quarries – at various sites around the Mediterranean, such as Malta, quite sophisticated layouts still exist where the early engineers carved out contour-following rutways, sidings and passing loops. Its limitation was that it could only be used where rock lay on or near the surface.

However, the evolutionary process of railways as we know them, with flanged wheels running on rails, began in the infancy of the Industrial Revolution in Europe. During the sixteenth century tub railways were used to manhandle coal and ore from mines in Germany, these crude forms of transportation running on wooden boards with the wooden wheeled tubs guided by a pin running in a slot between the boards. Archaeologists have discovered traces of such a system used by German miners in the Mines Royal near Keswick in the English Lake District.

The first recorded use of an overground wagonway anywhere in the world dates from 1604 when the 2-mile Wollaton Waggonway was opened for business in Nottinghamshire. With a gauge of 4 ft 6 in. it was engineered by Huntingdon Beaumont, an English mining engineer, to carry coal from mines in Strelley, owned by Sir Percival Willoughby, for onward shipment by barge on the River Trent. Horses were used to haul wooden wagons without flanged wheels along flanged wooden rails.

Huntingdon Beaumont also built wooden wagonways in Northumberland for transporting coal from his mines to Blyth harbour, and the end of the seventeenth century found an extensive network of horse-drawn wagonways in use in northeast England. By then the horse-drawn wagons were fitted with flanged wooden wheels running on oak rails laid on wooden ties, or sleepers, between collieries and staithes on the banks of the River Tyne in the Newcastle area and the River Tees in County Durham. It was a dangerous operation with loaded coal wagons running downhill by gravity on gradients as steep as 1 in 15,

progress being slowed by a brakeman operating an extremely primitive brake.

Early railway technology was problematic in that the wooden wheels quickly deteriorated, while cast-iron wheels, first introduced in 1734, soon wore out the wooden rails. To counter this iron straps were attached to the top of the rails but these became loose and it was soon apparent that iron rails were the only solution. The first wholly iron rails were manufactured by Richard Reynolds of Coalbrookdale in Shropshire in 1767, their introduction not only reducing repair costs but greatly increasing the loads hauled by horses due to reduced friction. At this time there was no standardization of gauge, which could vary from three to five feet.

The use of flanged wheels running on flangeless rails eventually became standard practice on Britain's railways, however two other systems were also introduced. In 1795 the English engineer Benjamin Outram used horse-drawn wagons with flangeless wheels that ran on L-shaped cast-iron rails for his 5-mile Little Eaton Gangway linking collieries to the newly opened Derby Canal. Known as 'plateways' these industrial canal-feeder railways flourished for a while, being particularly popular in South Wales – by the early nineteenth century there were nearly 150 miles of plateways operating along the valleys. However, this design was inherently weak as the inner upright flange was prone to breakage and the bottom horizontal plate collected dirt and stones thus impeding the wagons' progress. Despite its obsolescence and structural weaknesses Outram's original plateway in Derbyshire continued to be horse drawn until 1908.

PREVIOUS SPREAD: The Stockton & Darlington Railway, seen here on opening day in 1825, was the world's first public steam-operated railway. Engineered by George Stephenson, it was built to link collieries in West Durham and Darlington with the docks on the River Tees.

LEFT: Followed by a horse, a loaded wagon of coal runs under gravity on a wagonway in Newcastle, c.1773. In the background are coal staithes and keel boats that ferried coal to collier ships on the River Tyne.

BELOW: The Little Eaton Gangway opened in 1795 to link collieries with the Derby Canal. The horse-drawn wagons ran on L-shaped cast-iron rails and this early railway remained operational until the beginning of the twentieth century.

Benjamin Wyatt designed another system using double-flanged wheels running on oval rails for Lord Penrhyn's narrow-gauge slate quarry lines in North Wales. Introduced in 1801 it remained in use on Welsh quarry tramroads well into the twentieth century despite over-complicated trackwork at crossings and points.

By the late eighteenth century Britain's industrial revolution had gathered pace and soon there were hundreds of miles of primitive wagonways in use in Northumberland, County Durham, the Midlands and South Wales linking collieries, quarries, and copper- and iron-ore mines with rivers, canals and coastal harbours.

The late eighteenth and early nineteenth century was the Golden Age of canal building across Britain, this mode of transport seeming to be the answer to the country's transport needs for a short period leading up to the height of the Industrial Revolution. Horse-drawn wagonways, or canal tramroads, were just seen as useful feeders for the main canal network – but canal construction was a slow and costly business while wagonways were cheap and quick to build. Britain's canal network at its

peak in the early nineteenth century had expanded to nearly 4,000 miles but transporting raw materials and finished goods was a slow process with horse-drawn barges only moving at walking pace, often having to negotiate locks to gain or lose height. The 29-mile journey on the Worcester & Birmingham Canal, which opened in 1815, involved passing through fifty-eight locks and four single-lane tunnels, with journeys sometimes taking a couple of days.

Wagonways, tramroads and canals all depended on horsepower to keep them in business but waiting in the wings was one of the most important technological breakthroughs ever witnessed by mankind. Changing the face of the world forever, the Steam Age was about to revolutionize Britain's industries and its railways, in turn making it the most powerful nation on Earth for more than 100 years.

BELOW: Seen here in 1783, the Parkmore Waggonway was a simple horse-drawn railway that linked collieries around Gateshead with coal staithes on the River Tyne.

The first horse-drawn public railways

Before the advent of steam power Britain's coal-carrying wagonways and plateways depended entirely on horsepower to keep them moving. The vast majority were impermanent affairs usually laid on mine-owners' land or across private land requiring a wayleave from the owner, permitting the wagonway access in return for annual payments. These unsatisfactory arrangements ended following Parliamentary approval for new railways in which Acts of Parliament permitted companies to purchase land outright.

The latter years of the eighteenth and early years of the nineteenth centuries before the advent of steam power saw many world railway firsts in Britain.

Middleton Railway

Prior to 1758 all wagonways were built on an *ad hoc* basis but in that year the Middleton Railway of Leeds became the first railway in Britain to be granted powers of permanence by an Act of Parliament.

In the eighteenth century Middleton was an important coal-mining village near Leeds and coal had been mined there since the thirteenth century. Charles Branding, the owner of the mines of the Middleton Estates in the mid-eighteenth century, sought to transport coal from his mines to the nearby growing city of Leeds. Poor roads led him to build in 1755 a horse-drawn wagonway to the nearest river from where the coal was then transported to Leeds. In 1758 Branding obtained parliamentary approval to build a new wagonway – the Middleton Railway – by a more direct route from his mines to the River Aire in Leeds. The 4-ft 1-in.-gauge wagonway, just under one mile long, was an immediate success. In 1799 iron edge rails replaced the wooden rails and in 1812 the Middleton Railway became the world's first commercial railway to successfully employ steam locomotives. In 1881 the line was regauged to 4 ft 8½ in. and continued in commercial operation until 1960. Preservationists introduced passenger

services in 1969 and today this historic line is owned and operated by the Middleton Railway Trust.

Surrey Iron Railway

In 1803 the world's first railway company was established and the first publicly subscribed railway was opened throughout. Established by an 1801 Act of Parliament, the 9-mile Surrey Iron Railway was a horse-drawn 4-ft 2-in.-gauge double-track cast-iron plateway laid on stone blocks, running between Croydon and Wandsworth in what is now south London. Built instead of an originally planned canal, the railway charged tolls to private users in similar fashion to the method employed on our modern railways. The mainstay of this horse-drawn freight-carrying operation was the carriage of coal, agricultural products, stone and timber. Following success in its early years the railway fell on hard times after the opening of the Croydon Canal in 1811, only just keeping its financial head above water until closure in 1846. Today parts of the route are used by the London Tramlink.

LEFT: Opened between Wansdworth and Croydon in 1803, the Surrey Iron Railway was the world's first publicly subscribed railway. It was a freight-only line with wooden wagons being horse drawn along a cast-iron plateway.

BELOW: The railway charged tolls to private users but closed in 1846 following the opening of the Croydon Canal.

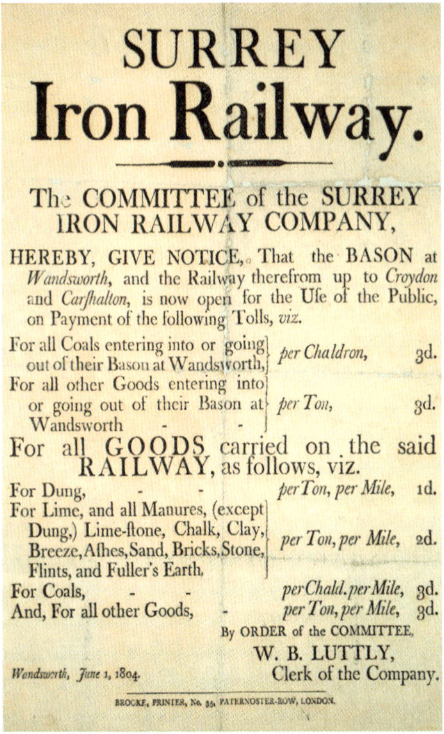

SURREY
Iron Railway.

The COMMITTEE of the SURREY IRON RAILWAY COMPANY,

HEREBY, GIVE NOTICE, That the BASON at *Wandsworth*, and the Railway therefrom up to *Croydon* and *Carſhalton*, is now open for the Uſe of the Public, on Payment of the following Tolls, *viz.*

For all Coals entering into or going out of their Bason at Wandsworth,	*per Chaldron,*	3d.
For all other Goods entering into or going out of their Bason at Wandsworth	*per Ton,*	3d.

For all GOODS carried on the said RAILWAY, as follows, viz.

For Dung,	*per Ton, per Mile,*	1d.
For Lime, and all Manures, (except Dung,) Lime-ſtone, Chalk, Clay, Breeze, Aſhes, Sand, Bricks, Stone, Flints, and Fuller's Earth,	*per Ton, per Mile,*	2d.
For Coals,	*per Chald. per Mile,*	3d.
And, For all other Goods,	*per Ton, per Mile,*	3d.

By ORDER of the COMMITTEE,
W. B. LUTTLY,
Clerk of the Company.

Wandsworth, June 1, 1804.

BROCKE, PRINTER, No. 35, PATERNOSTER-ROW, LONDON.

Oystermouth Railway

The world's first passenger-carrying public railway was also horse drawn. Following Parliamentary approval, the 4-ft-gauge Oystermouth Railway (later known as the Swansea & Mumbles Railway) first carried passengers in 1807. The 5½-mile plateway ran around Swansea Bay between Swansea and Oystermouth. From the world's first railway station, the Mount terminus in Swansea, passengers were carried in horse-drawn stagecoach-style vehicles. Slightly ahead of its time, passenger services ceased when a new turnpike road was opened in the 1820s! The railway then became derelict until being reopened as a standard-gauge horse-drawn coal-carrying line in 1842. The horse-drawn passenger service restarted in 1855, with the introduction of steam power in 1877.

The final chapter in the story of this historic railway came in 1929 when it was electrified using overhead power equipment with passenger services being provided by large double-deck tram cars built by Brush of Loughborough. The railway was purchased by the operator of the local bus company in 1958 and, unsurprisingly, its new owner abruptly closed the line two years later.

Kilmarnock & Troon Railway

The 10-mile Kilmarnock & Troon Railway (K&TR) was built to transport coal from collieries around Kilmarnock to the harbour at Troon, a journey previously undertaken by horse and cart along poorly made roads. In 1808 it became the first railway in Scotland to be authorized by an Act of Parliament. Engineered by William Jessop who had previously built the Surrey Iron Railway in England, the K&TR was also a double-track toll plateway with wrought-iron L-shaped rails but with a gauge of 4 ft. Opened for coal traffic in 1812, a privately run passenger service in a converted stagecoach called the *Caledonia* was introduced in 1813.

In 1817 the K&TR became the first railway in Scotland to operate a steam locomotive. Supplied by George Stephenson, it was based on his locomotives built for the Killingworth Colliery near Newcastle-upon-Tyne, but while these had flanged wheels operating on edge rails, the K&TR loco had flangeless wheels for working on the L-shaped plateway. However, although it was capable of hauling ten tons of coal at 5 mph it was soon replaced by horsepower as its 5-ton weight proved too much for the constantly breaking brittle plateway.

The shortcomings of plateways eventually led the K&TR to apply to Parliament for authority to convert their railway to steam locomotive operation. Following Parliamentary approval in 1837 the line was re-laid with edge rails to the standard gauge of 4 ft 8½ in. Horsepower continued until 1841, when steam locomotives were introduced. The section of the original route between Kilmarnock and Barassie is today used by ScotRail passenger trains as well as seeing the passage of coal traffic.

RIGHT: Similar to early operations on the Kilmarnock & Troon Railway, this horse and dandy cart operated passenger services on the North British Railway between Port Carlisle and Drumburgh until 1914.

The birth of the Steam Age

While the coal-carrying wagonways continued to depend on horsepower to keep them operating, the first tentative steps to harness the power of steam were being made in southwest England to pump water out of mines. First on the scene was Thomas Savery of Devon who patented an atmospheric steam engine in 1698 – apart from some taps it had no working parts and in practice it was unable to pump water from more than a depth of thirty feet. Thomas Newcomen, also from Devon, is credited with designing the world's first practical steam pumping engine; it was the first to operate using a cylinder and moveable piston and was capable of raising water from a depth of 150 feet. Introduced in 1710 Newcomen's pumping engines were soon in use throughout Britain and Europe, some remaining in operation for over 150 years.

A Scottish-born engineer, James Watt, went on to improve on Newcomen's design and in 1775, in partnership with Birmingham foundry owner Matthew Boulton, introduced his world-beating low-pressure stationary steam engine. It was an instant success and hundreds were sold in kit form to mine owners around the world – Boulton & Watt's engineers reassembled them in situ. Although initially used to pump water from mines, Watt's engines had many other industrial applications, playing a leading role in the rapid growth of Britain's Industrial Revolution.

Additional impetus came when, in the latter part of the eighteenth century, the first crude steam-powered road vehicles appeared. The first of these was William Murdoch's high-pressure miniature steam carriage which was successfully demonstrated in a Cornish hotel in 1795, making it the first working steam locomotive in Britain. Also in Cornwall Richard Trevithick went on to develop the use of high-pressure steam and demonstrated a full-size steam road passenger-carrying carriage on the streets of Camborne in 1801. The *Puffing Devil's* range was limited as it soon ran out of steam but this was the first time in

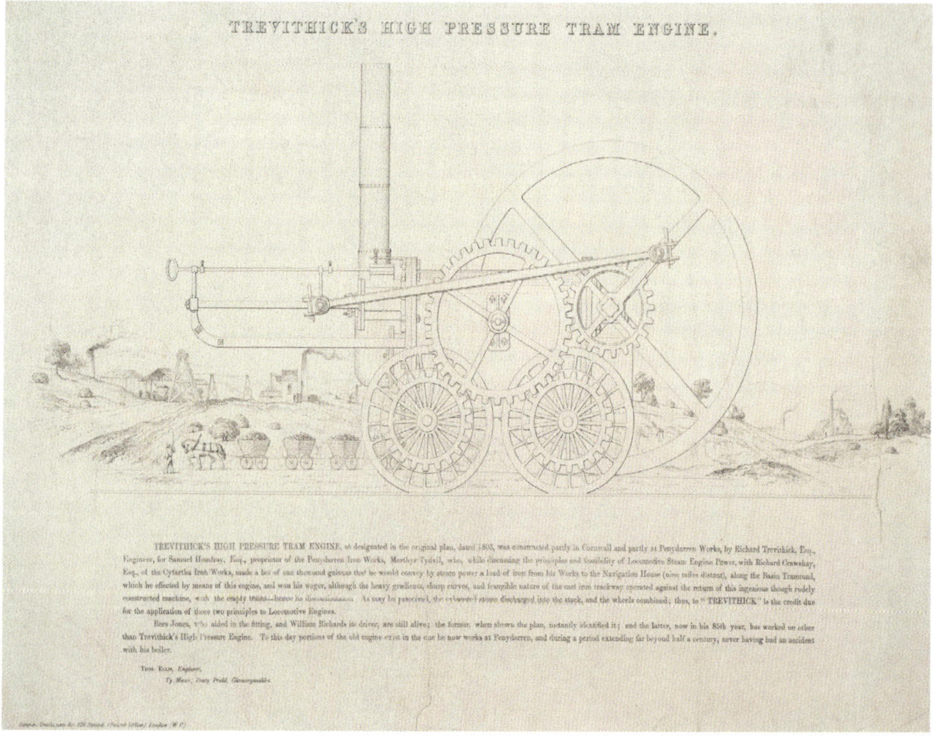

the world that passengers had been transported by a steam vehicle. Trevithick continued with his development of high-pressure steam engines and was commissioned to build a steam railway locomotive for the 9-mile, 4-ft 2-in.-gauge Penydarren plateway, serving ironworks at Dowlais in South Wales. Despite the locomotive's successful maiden voyage with a loaded train in 1804 – it hauled ten tons of iron and seventy men in five wagons a distance of nine miles at an average speed of 2 mph – the cast iron plate rails broke under its weight and the experiment ended there. However, the viability of steam haulage was proved beyond a shadow of doubt, although it was to be some years before it superseded traditional horsepower. In 1805 a second Trevithick steam locomotive with flanged wheels was built for the Wylam Colliery near Newcastle but this also proved too heavy for the wooden wagonway on which it ran.

With the Napoleonic Wars seeing ever-increasing prices for horse fodder, the next ten years saw major strides forward in the development of the steam locomotive. Trevithick whetted the public's appetite when he ran his circular demonstration steam-hauled train in Bloomsbury (London) in 1808. Hauled by his third steam railway locomotive, *Catch Me Who Can*, it gave rides to the public but was not a great success and he died penniless in 1833, despite his fame as the builder of the world's first railway locomotive.

LEFT: Richard Trevithick's high-pressure steam locomotive became the first in the world to successfully run on rails. It was used at the Penydarren Iron Works in Merthyr Tydfil in 1803.

BELOW: John Blenkinsop's rack locomotive was successfully introduced in 1812 to haul coal wagons on the Middleton Railway near Leeds.

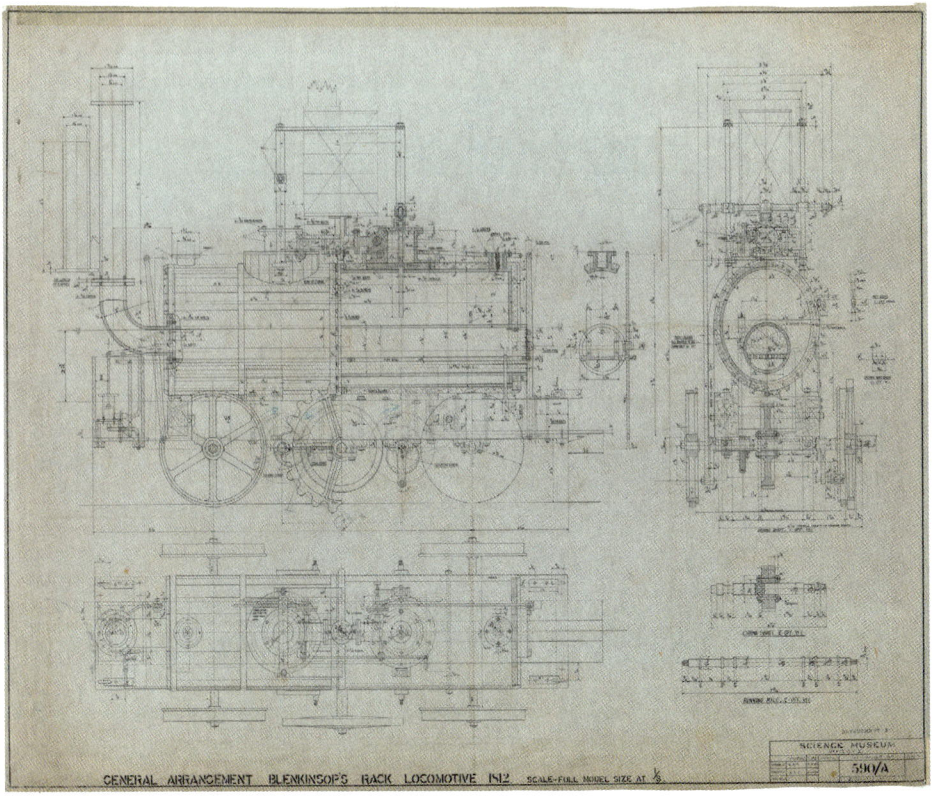

In Yorkshire, John Blenkinsop designed two steam locomotives operating with cogs and rail-mounted racks ('rack and pinion') for use on the privately owned Middleton colliery railway near Leeds. Built by Matthew Murray of Holbeck, Leeds, these first ran in 1812 and continued in use for thirty years on the world's first commercially successful steam railway.

The Wylam colliery railway near Newcastle had already been operating as a wooden wagonway for nearly 200 years. Having witnessed Trevithick's locomotive in 1805, the colliery's chief engineer William Hedley and his foreman Timothy Hackworth had the 5-mile line converted to a 5-ft-gauge plateway in 1813 and successfully introduced their new steam locomotive, *Puffing Billy*. Building on Trevithick's proven design of an efficient high-pressure boiler with a single straight flue, Hedley's adhesion loco initially proved too heavy for the plate rails. To overcome this problem he successfully rebuilt it as an articulated loco, spreading the weight with two 4-wheel bogies driven by gears. Along with a sister engine named *Wylam Dilly* the locomotive operated until 1830 when it was converted into a four-wheeler with flanged wheels to allow it to run on newly laid iron-edge track. Both locomotives continued in operation in this form until 1862. Fortunately for posterity both of the world's oldest surviving steam locomotives have been preserved – *Puffing Billy* at the Science Museum in London and *Wylam Dilly* in Edinburgh's Royal Museum.

At the same time as Hedley's and Hackworth's success at Wylam a local self-taught engineer was about to launch himself onto an unsuspecting world. Born in Wylam in 1781, George Stephenson was soon to be known as the 'Father of Railways', the first great engineer of the Railway Age.

LEFT: Seen here in 1876, William Hedley's *Puffing Billy* was successfully introduced in 1813 to haul coal wagons at Wylam Colliery near Newcastle-upon-Tyne. Now on display at the Science Museum in London, it is the world's oldest surviving steam railway locomotive.

George Stephenson

George Stephenson was born of illiterate parents in Wylam, Northumberland in 1781, his father working as a fireman for the Wylam Colliery stationary pumping engine. Opened in 1748 to transport coal from the colliery down to the River Tyne the Wylam Waggonway passed close to George's home and as a young man he would have seen his first steam locomotive – in 1805 Richard Trevithick's second steam locomotive was briefly used on the line but proved too heavy for the wooden rails. George's first job was herding cows but at the age of 14 he went to work with his father at the colliery, learning to control the steam-powered winding gear. Intent on improving himself George also attended night classes to learn the 3 Rs.

In 1802 George was promoted to colliery engineman and later that year married local girl Francis Henderson. Their son, Robert (later to become an important railway engineer in his own right), was born a year later but a daughter Fanny died within months of her birth in 1805. George's sorrows continued when Francis died of TB in 1806. After being appointed as engineman at Killingworth Colliery in 1808, he would take apart and put back together the stationary steam engines on his days off. The colliery owners recognized his growing knowledge of steam engines and appointed him as enginewright at Killingworth in 1812, where he was responsible for maintaining and repairing the colliery's stationary steam engines.

Following Hedley and Hackworth's successful introduction of their locomotive, *Puffing Billy*, at Wylam Colliery in 1813 George went on to build his first steam locomotive. Named *Blücher* after the Prussian general at the Battle of Waterloo, it was built in a colliery workshop behind his house in 1814 and in many respects was similar to Blenkinsop's machines on the Middleton Railway, with the notable difference that its flanged wheels ran on normal edge rails without the use of rack-and-pinion. It was a great success as the first practical locomotive of its type in the world, hauling thirty tons of coal uphill at 4 mph. George went on to build a number of these locomotives for use at Killingworth, Hetton Colliery and the Kilmarnock & Troon Railway – the latter was six-wheeled but proved too heavy for the cast-iron rails.

In 1820, the same year that he married his second wife Betty, Stephenson was appointed to build a railway linking Hetton Colliery and staithes on the River Wear in County Durham. A combination of gravity inclines and

locomotive-hauled sections, the 8-mile line became the first railway in the world not to use horsepower. Built to a gauge of 4 ft 8 in. – the same that George had used at Killingworth – it was opened in 1822 and remained operational until 1959.

George also invented a miners' safety lamp that was widely used in northeastern pits although Humphrey Davy claimed that Stephenson had stolen his own invention – in the end George was exonerated by a House of Commons committee and found to have equal claim for its invention. He also patented an overlapping joint for joining cast-iron rails. Early rails were short 3-ft lengths and made from cast iron, but were brittle and broke easily, especially under the weight of a steam locomotive. A breakthrough came when, in 1820, Northumberland railway engineer John Birkinshaw

patented 15-ft-long wrought-iron rails and it was these that Stephenson used for the building of the Stockton & Darlington Railway instead of his own patented cast-iron rails. Birkinshaw went on to become Robert Stephenson's assistant engineer during the building of the London & Birmingham Railway.

ABOVE: George Stephenson engineered the steam-operated railway at Hetton Colliery in County Durham. His locomotives were capable of hauling a 64-ton train at 4 mph.

LEFT: George Stephenson (1781–1848), the 'Father of Railways'.

George Stephenson's greatest triumphs were yet to come. He was appointed Chief Engineer of the Stockton & Darlington Railway in 1822 and opened a locomotive works in Newcastle jointly with S&DR director Edward Pease, installing his son Robert as managing director. After his success with the S&DR – on opening in 1825 it was the world's first public railway to use steam locomotives – Stephenson went on to engineer the world's first inter-city railway, the Liverpool & Manchester Railway, which opened in 1830.

George Stephenson was consequently inundated with requests from railway promoters – the early American railroads ordered their first locomotives from his factory in Newcastle. He worked with his former assistant Joseph Locke as civil engineer on the building of the Grand Junction Railway and was chief engineer for several English railways including the Manchester & Leeds, the York & North Midland and the North Midland. He discovered rich seams of coal in Derbyshire while tunnelling for the latter company and went into business with railway promoters George Hudson and Joseph Sanders, opening coal mines, ironworks and limestone quarries around Chesterfield. Stephenson spent his latter years living in nearby Tapton House, running a small farm where he experimented in stock breeding and the speeding up of fattening chickens. One of his quirkier inventions was a straight glass tube for growing straight cucumbers, an apparatus that no doubt would have been embraced wholeheartedly by the European Union if it had existed at that time!

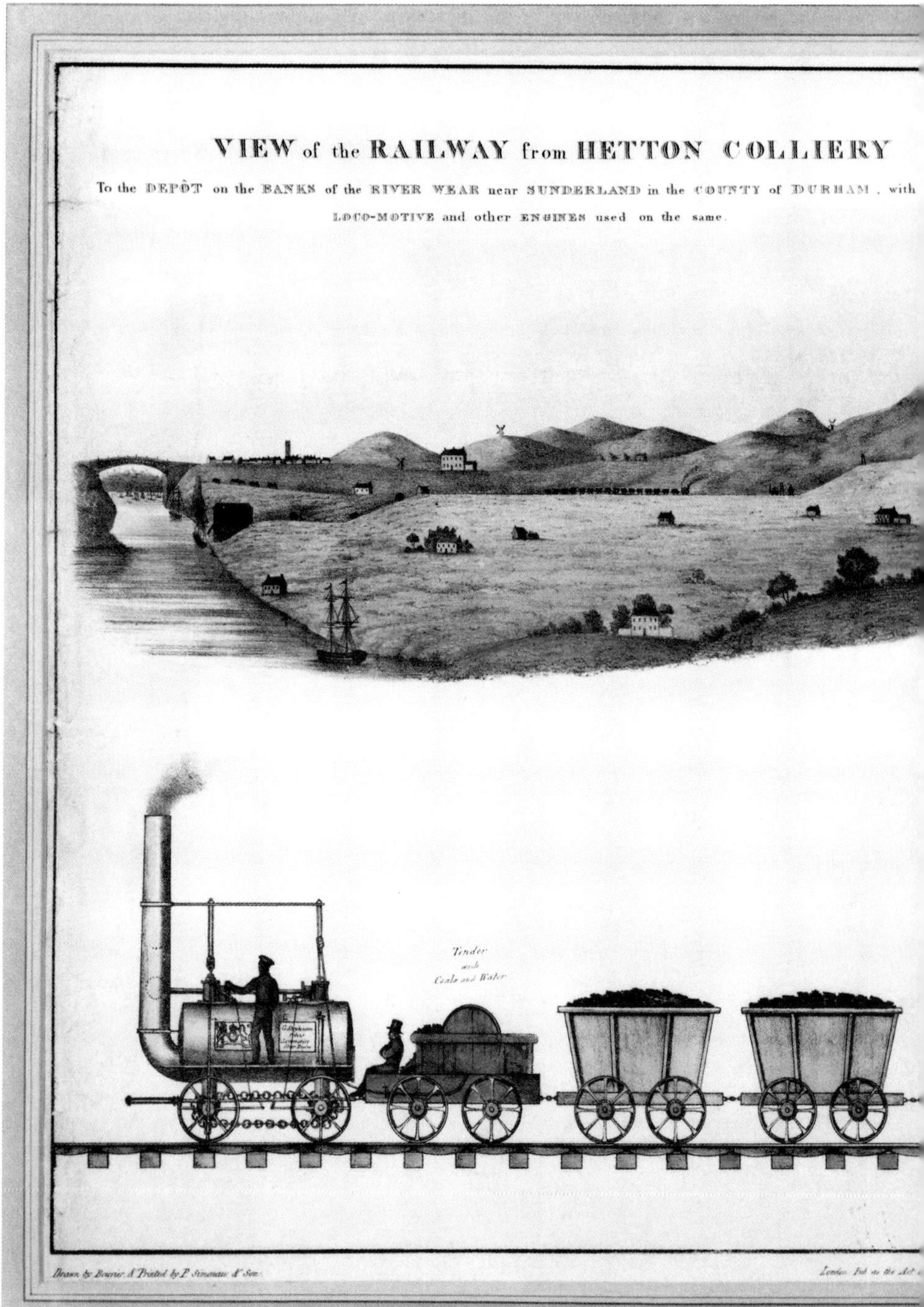
VIEW of the RAILWAY from HETTON COLLIERY
To the DEPÔT on the BANKS of the RIVER WEAR near SUNDERLAND in the COUNTY of DURHAM, with
LOCO-MOTIVE and other ENGINES used on the same.
Tender
with
Coals and Water
Drawn by Bourne, & Printed by P. Simonau & Son.

A view of the railway engineered by George Stephenson at Hetton Colliery in County Durham. Opened in 1822, the colliery was the home of the first extensive private railway in the world, using both stationary steam engines and locomotives to transport coal to the banks of the River Wear, near Sunderland.

Stockton & Darlington Railway

George Stephenson was appointed Chief Engineer of the Stockton & Darlington Railway in 1822, following the success of his early colliery railways and steam locomotives.

During the eighteenth century and early years of the nineteenth century coal from mines in County Durham was transported by horse and cart to the nearest navigable river or coastal harbour. Proposals for improving navigation on the River Tees came to nothing, but a horse-drawn tramroad with inclined planes received more support, especially from local Quaker businessmen for whom Edward Pease was the main driving force. Although several surveys were carried out for the tramroad, each time the plans were laid before Parliament they were opposed by local landowners. After changing the route yet again an Act of Parliament was finally passed in 1821 allowing the building of the Stockton & Darlington Railway (S&DR) between inland collieries at Witton Park near Shildon to the River Tees at Stockton via Darlington. However, the Act stipulated that it was to be a toll railway using horsepower – steam locomotives apparently did not exist!

Despite the new railway receiving Parliamentary approval, Edward Pease was dissatisfied with the planned route of the line and appointed George Stephenson to carry out a new survey. Aided by his 18-year-old son Robert, George reported back that a shorter route laid with the new wrought-iron edge rails patented by John Birkinshaw would be preferable to the route approved by Parliament, and strongly arguing the case for using steam locomotives.

In 1822 George Stephenson was appointed Chief Engineer of the S&DR and a year later received Parliamentary approval for his new 25-mile route, the use of 'moveable engines' and the carrying of passengers. A new locomotive factory in Newcastle-upon-Tyne was opened in the same year by George Stephenson and Edward Pease, with young Robert in place as managing director – orders for two steam locomotives were soon placed by the S&DR. Robert Stephenson & Company was now in business and it would play an important role in supplying steam locomotives for railways around the world until the 1950s.

The new railway was to be built to the gauge of 4 ft 8 in. which Stephenson had already used for his colliery railways at Killingworth and Hetton. With an extra half inch added to reduce friction on curves, this was soon adopted as the standard gauge for the majority of railways around the world and remains so today.

ABOVE: Built at Robert Stephenson's locomotive factory in Gateshead, 0-4-0 locomotive *Locomotion* hauled the first train on the Stockton & Darlington Railway in 1825. It is now on display at the Darlington Railway Museum in County Durham.

BELOW: Seen here shortly after opening in 1825, a loaded coal train crosses the suspension bridge carrying the Stockton & Darlington Railway over the River Tees near Stockton. The 4-ft 8½-in.-rail gauge adopted by George Stephenson has since become the standard gauge for over 60 per cent of the world's railways.

While the western end of the line was operated by inclines with stationary steam engines (also supplied by Robert Stephenson & Co), the twenty miles from Shildon to Stockton Quay were to be operated by steam locomotives. The new line stretched the technology of the time with embankments of nearly 50 feet in height, a wrought-iron girder bridge across the River Gaunless and a graceful arched stone bridge over the River Skearne. The actual cost of building the railway far exceeded Stephenson's original estimate, an unfortunate trait that was to dog him on future railway projects.

The opening day of the Stockton & Darlington Railway on 27 September 1825 was as important a date in world history as 20 July 1969, when man first walked on the Moon. On this day, watched by thousands of people, George Stephenson's *Locomotion* hauled a passenger-carrying steam train on a public railway for the first time in the world. Around 500 passengers on this first train were carried in the railway's only passenger coach (*The Experiment*) and in twenty-one converted coal wagons. Built at Stephenson's new locomotive works in Newcastle, *Locomotion* was the first to feature driving wheels connected by coupling rods instead of gears. T-section wrought-iron rails, connected by 'fishplates' at 15-ft intervals, were laid on sturdy stone blocks designed to take the weight of heavily loaded coal trains.

A crowd of 40,000 and a 21-gun salute greeted the world's first steam-hauled passenger train upon arrival at Stockton Quay! The Railway Age had arrived.

The immediate success of the new line finally proved the viability of steam railways. Apart from the inaugural run on opening day, however, passengers were still carried in horse-drawn coaches until the introduction of steam-hauled services in 1833. *Locomotion* was retired in 1841 when it became a stationary steam engine until being preserved in 1857. It was moved to the Darlington Railway Centre and Museum in 1975 following many years on display at Darlington Bank Top station.

In later years the Stockton & Darlington Railway expanded eastwards to Middlesbrough Docks and went on to transport iron ore from the Cleveland Hills to blast furnaces alongside the River Tees as well as being a major coal haulier. It further expanded westwards with takeovers in 1860 of railways from Barnard Castle to Penrith and Tebay on the West Coast Main Line. With its 200 route miles the company was taken over by the North Eastern Railway in 1863, while today the majority of its original 1825 route is still open for business, served by passenger trains on the Tees Valley Line.

LEFT: This map of the Stockton & Darlington Railway shows the original route surveyed in 1821 that envisaged the use of horsepower and inclined planes with users paying tolls.

BELOW: The 150th anniversary of the opening of the Stockton & Darlington Railway was celebrated in this painting by Terence Cuneo.

1826	1827	1828	1829	1830	1831	1832	1833	1834
Liverpool & Manchester Railway receives Royal Assent			Robert Stephenson's *Rocket* wins the Rainhill Trials	The world's first inter-city railway, the Liverpool & Manchester Railway, opens	Innocent Railway opens in Edinburgh		Brunel appointed Chief Engineer of Great Western Railway	

THE FIRST INTERCITY and TRUNK RAILWAYS

1826–1842

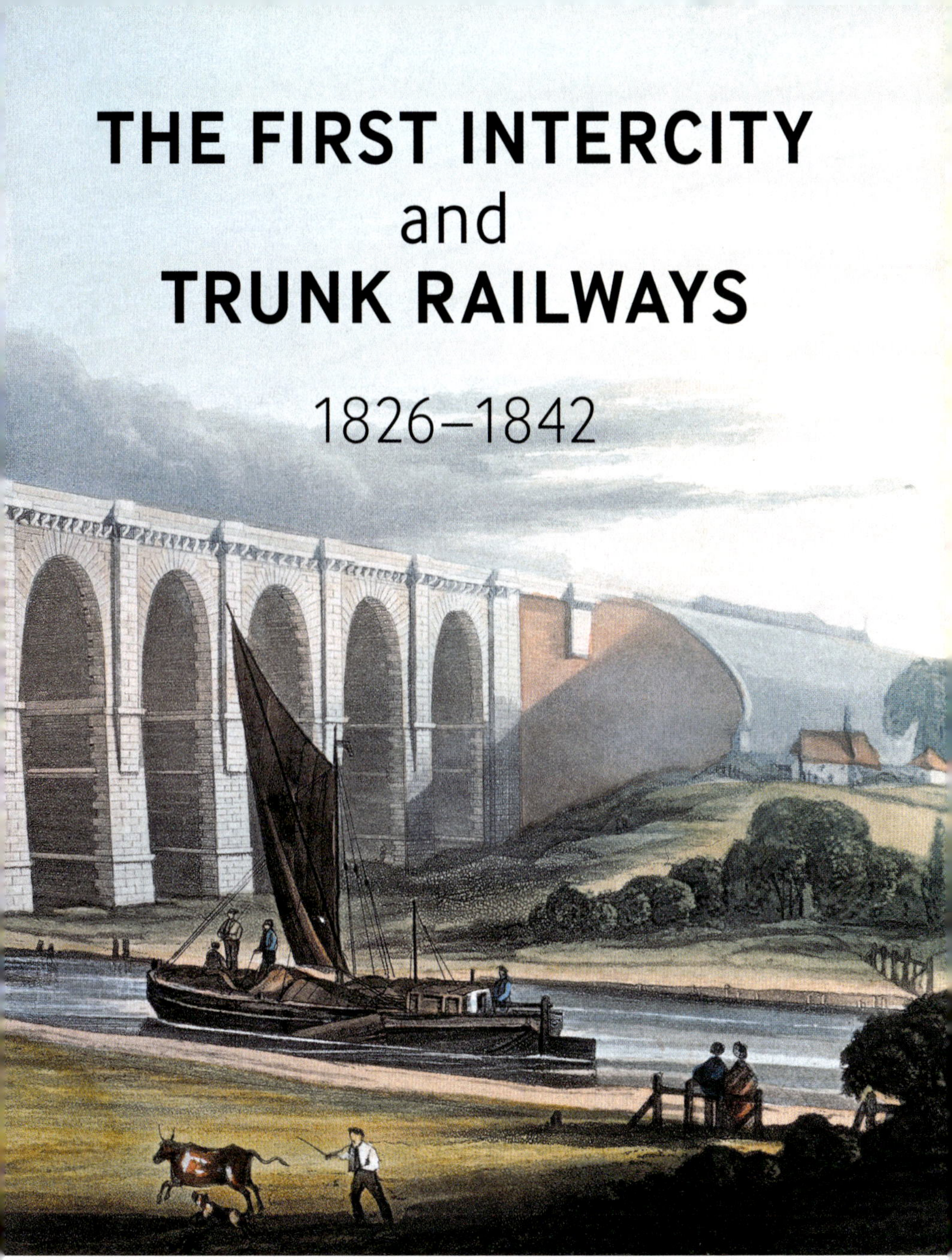

1835	1836	1837	1838	1839	1840	1841	1842
Horse-drawn Edinburgh and Dalkeith Railway opens	Horse-drawn Ffestiniog Railway in North Wales opens	Grand Junction Railway opens	London & Birmingham Railway opens	World's first railway timetable published by George Bradshaw	Thomas Cook organizes his first rail excursion train	Broad-gauge Great Western Railway opens between Paddington and Bristol	Queen Victoria makes first royal train journey

The Rainhill Trials

While the Liverpool & Manchester Railway (L&MR) was being built the company's directors organized a competition to find out if locomotives were more suitable for hauling their trains than stationary steam engines. At this early stage in their development steam locomotives were in their infancy and often proved unreliable – even until 1833 the Stockton & Darlington Railway continued to rely on horsepower to haul their passenger trains.

The competition took place over six days in October 1829 at Rainhill on a 1½-mile section of level track that had already been completed. The company offered a prize of £500 to the winner. Performing in front of a panel of distinguished officials, each locomotive had to satisfy the rather complicated conditions on weight and load set by the organizers and were required to make ten return trips without refuelling, equivalent to travelling the thirty-five miles between Liverpool and Manchester, at an average speed of not less than 10 mph with thirty miles performed at top speed. After this gruelling test the locomotives then had to repeat the process, equivalent to a return journey from Manchester to Liverpool.

The organizers set a severe weight limit – 4½ tons for a 4-wheeled locomotive or six tons for a 6-wheeled locomotive – this effectively barred any existing locomotives already in service, such as the 8-ton *Locomotion* on the Stockton & Darlington Railway.

The competition caught the imagination of the British public and attracted ten entries, although five of these, such as a perpetual motion machine, were purely flights of fancy existing only on paper. The five entries that succeeded in starting the competition were themselves a very mixed bunch and all but one fell by the wayside as the competition progressed.

First to withdraw was a wacky horse-powered machine called *Cycloped*. Designed by Thomas Brandreth of Liverpool, who was also a director of the L&MR, it was powered by a horse walking on a continuous treadmill which soon became unstuck when the horse fell through the floor. The next unsuccessful candidate was Timothy Burstall's steam locomotive, *Perseverance,* which was damaged en route to Rainhill from Scotland. Burstall managed to repair it in time for its trials on the sixth and final day but it could only achieve 6 mph – despite this he was awarded a consolation prize by the organizers for his endeavours in the face of adversity.

The third locomotive to withdraw was Timothy Hackworth's *Sans Pareil*. While a more serious competitor, it was overweight and its steam technology was already obsolete. Employing vertical cylinders driving two pairs of connected driving wheels, it lacked the more modern fire-tube boiler of the eventual winner, Stephenson's

Rocket. It consumed prodigious amounts of fuel and eventually had to be withdrawn due to a cracked cylinder. Despite these failures the L&MR bought the locomotive and after several careers it ended up as a stationary boiler before being presented to the Patent Office Museum in 1864. This historic locomotive is currently on display at the Shildon Locomotion Museum in Co Durham.

The penultimate steam competitor was a bit of a lightweight. Built by John Ericsson and John Braithwaite – better known for their horse-drawn fire engines fitted with steam pumps – the *Novelty* was an 0-2-2WT (i.e. carrying its water supply in a well between the wheels) capable of raising steam very quickly and incorporating some novel features including a copper firebox and mechanical blower. Fitted with two vertical cylinders it was also fast and achieved an astonishing 28 mph on the first day of the competition. Unfortunately the locomotive then experienced several failures with its blower and water feed pipe and was forced to withdraw.

The ultimate winner of the Rainhill Trials was Robert Stephenson's 0-2-2 *Rocket*. Built by his locomotive building company in Newcastle it featured many ground-breaking innovations that kick-started Britain's railway revolution and the evolution of steam locomotive technology. This winning locomotive combined a single pair of driving wheels that provided much greater adhesion than a 0-4-0 type. They were driven directly by a pair of pistons in angled cylinders, unlike the unwieldy upright cylinders employed in earlier locomotives. The *Rocket* was also fitted with a separate firebox, a multi-tube

PREVIOUS SPREAD: A train crosses the viaduct over the Sankey Valley near Warrington on the Liverpool & Manchester Railway in February 1831. The viaduct has nine arches, each spanning 50 ft, and cost the company £45,000 to build. It has stood the test of time and is still in use today.

LEFT: Although Timothy Burstall's locomotive *Perseverance* was damaged en route to Rainhill, its builder was awarded a consolation prize by the organizers for his endeavours in the face of adversity.

RIGHT: The remains of Timothy Hackworth's locomotive *Sans Pareil* can be seen today at Shildon Railway Museum. Although it performed well at Rainhill the locomotive was excluded from the prize money of £500 because it was over the permitted weight limit.

BELOW: Headed by the winning entry, Robert Stephenson's *Rocket*, contestants parade in front of judges and spectators at the Rainhill Trials in Lancashire.

boiler and a blastpipe, all resulting in significantly reduced fuel consumption.

The *Rocket* was the only locomotive to successfully complete the trials, averaging 12 mph while hauling a load of thirteen tons at a top speed of 29.1 mph. It was declared the winner and went on to work on the L&MR between 1830 and 1834. Robert Stephenson was also contracted to build further locomotives for the railway including the *Northumbrian*, which was fitted with horizontal cylinders and drew the inaugural train on opening day in 1830. Later locomotives such as the 2-2-2s *Planet* and *Patentee* were fitted with horizontal inside cylinders mounted between the frames at the front – Robert Stephenson & Co were exporting the latter type to Germany, Holland, Belgium, Italy and Russia by the late 1830s. By this time Britain led the world in steam locomotive design.

After retiring from the L&MR in 1834 the *Rocket* went on to more humble duties working on a colliery railway in Cumberland before being presented to the Patent Office Museum in 1862. It is now preserved in its later modified state at the Science Museum in London.

LEFT: Much modified since winning the Rainhill Trials, the remains of Robert Stephenson's *Rocket* can be seen today at the Science Museum in London.

BELOW: Side and front elevations of the *Rocket* locomotive. It worked on the Liverpool & Manchester Railway until 1836 before moving to a colliery railway in County Durham. Although preserved in 1862 its tender and other parts are missing.

RIGHT: Details of the performance of *Rocket* at the Rainhill Trials were recorded in the notebook of one of the judges, John Urpeth Rastrick, a locomotive engineer from Stourbridge.

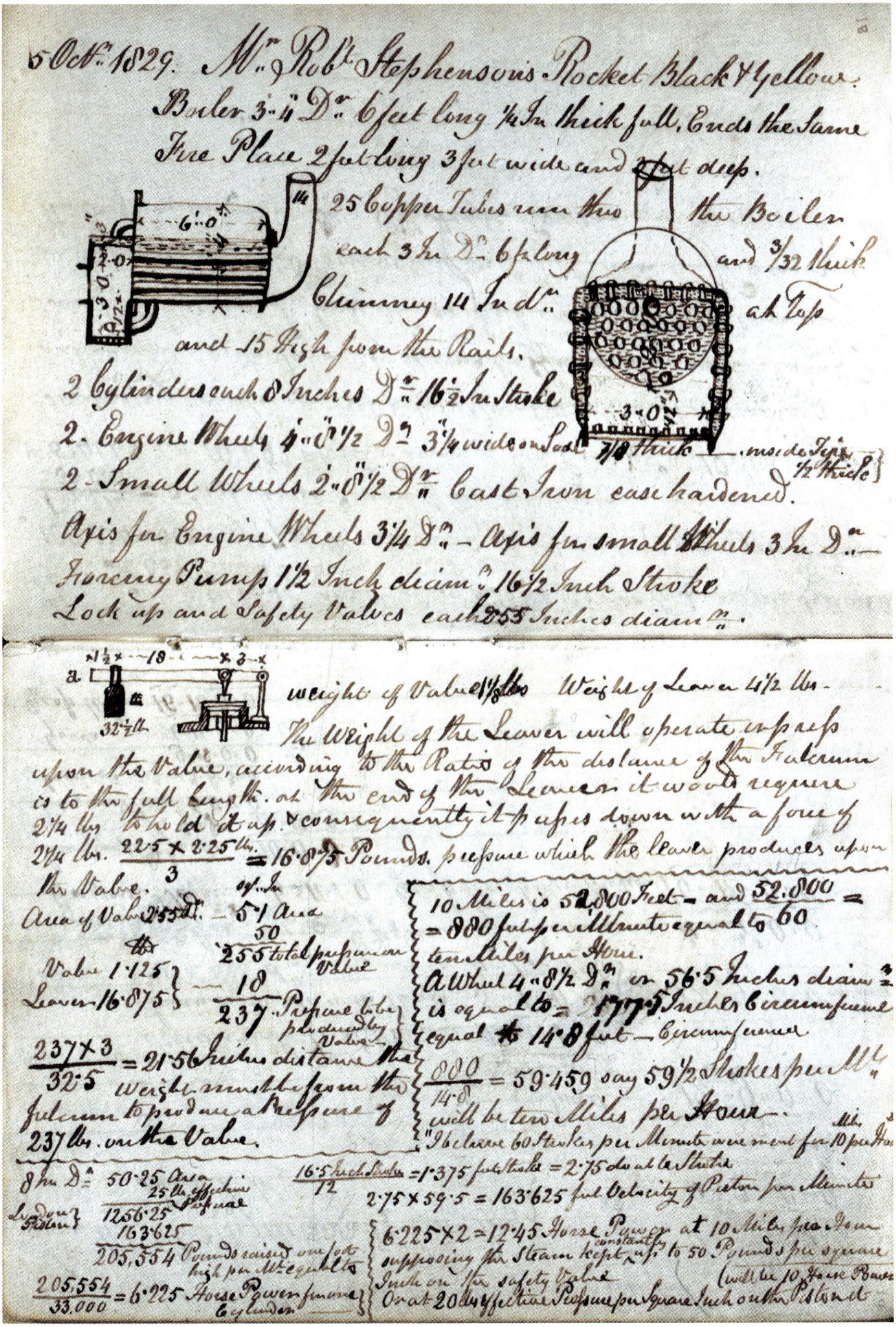
5 Octr. 1829. Mr. Robt. Stephenson's Rocket Black & Yellow.
Boiler 3-4 Dr. 6 feet long ¼ In thick full. Ends the Same
Fire Place 2 feet long 3 feet wide and 3 feet deep.
25 Copper Tubes run thro the Boiler
each 3 In Dr. 6 ft long and 3/32 thick
Chimney 14 In dr. at Top
and 15 High from the Rails.
2 Cylinders each 8 Inches Dr. 16½ In Stroke
2 - Engine Wheels 4-8½ Dr. 3¼ wide on Sole
2 - Small Wheels 2-8½ Dr. Cast Iron case hardened.
Axis for Engine Wheels 3¼ Dr. - Axis for small Wheels 3 In Dr. -
Forcing Pump 1½ Inch diamr. 16½ Inch Stroke
Lock up and Safety Valves each 2.55 Inches diamr.

weight of Valve 1⅛ lbs Weight of Leaver 4½ lbs -
The Weight of the Leaver will operate as press upon the Valve, according to the Ratio of the distance of the Fulcrum is to the full length. at the end of the Leaver it would require 2¼ lbs to hold it up. & consequently it presses down with a force of 2¼ lbs. 22.5 × 2.25 lbs / 3 = 16.875 Pounds pressure which the leaver produces upon the Valve.

Area of Valve 2.55 dr. - 5.1 Area
50
255 total pressure on Valve
Valve 1.125
Leaver 16.875 } - 18
237 Pressure to be produced by Valve
237 × 3 / 32.5 = 21.56 Inches distance the Weight must be from the fulcrum to produce a Pressure of 237 lbs. on the Valve.

10 Miles is 52,800 Feet - and 52,800 / 60 = 880 feet per Minute equal to ten Miles per Hour.
A Wheel 4-8½ Dr. or 56.5 Inches diamr. is equal to 177.5 Inches Circumference equal to 14.8 feet - Circumference
880 / 14.8 = 59.459 say 59½ Strokes per Mt. will be ten Miles per Hour.
I believe 60 Strokes per Minute were meant for 10 per Hour

8 In Dr. 50.25 Area
25 lbs effective Pressure
1256.25
163.625
205,554 Pounds raised one foot high per Mt. equal to
205,554 / 33,000 = 6.225 Horse Power from one Cylinder

16.5 Inch Stroke / 12 = 1.375 feet Stroke = 2.75 double Stroke
2.75 × 59.5 = 163.625 feet Velocity of Piston per Minute
6.225 × 2 = 12.45 Horse Power at 10 Miles per Hour supposing the Steam kept constantly up to 50 Pounds per square Inch on the safety Valve (will be 10 Horse Power)
Or at 20 lbs effective Pressure per Square Inch on the Piston.

Liverpool & Manchester Railway

Even before the opening of the Stockton & Darlington Railway a group of wealthy Lancashire businessmen had founded the Liverpool & Manchester Railway Company. The railway was surveyed by George Stephenson in 1824 and designed to provide cheap transport for raw materials and finished goods between the port of Liverpool and the textile mills of east Lancashire. With his son, Robert, in South America, George relied too much on his subordinates to check his calculations and inaccuracies in the survey resulted in the first Bill being rejected by Parliament in 1825. Stephenson was sacked and replaced by George and John Rennie as engineers and Charles Vignoles as surveyor. This time they got it right and the second Bill received Royal Assent in 1826, despite much objection from canal and landowners.

The Rennies' cost of building the railway proved too expensive, however, so George Stephenson was reappointed as engineer with Joseph Locke as his second-in-command. The 35-mile double-track line, built to a gauge of 4 ft 8½ in., was heavily engineered with sixty-four bridges and viaducts, a 1-mile 490-yd tunnel under Liverpool (the first railway tunnel in the world to run beneath a city), the 2-mile-long Olive Mount rock cutting and the crossing of Chat Moss on a raft of wood and heather hurdles. It was a triumph of engineering and stands the test of time to this day.

While construction of the line was proceeding George's son Robert was also developing a steam locomotive, the *Rocket*, in preparation for a competition to be held at Rainhill in October 1829. Competing against three other steam locomotives and a horse-powered version, Robert Stephenson's *Rocket* won hands down, scooping the £500 prize, setting a speed record of 29.1 mph and being awarded a contract to construct locomotives for the railway.

On 15 September 1830 the Liverpool & Manchester Railway opened to great celebrations. The inaugural train ran from Liverpool to Manchester and was hauled

by Robert Stephenson's locomotive, *Northumbrian*. The train conveyed the Duke of Wellington, then Prime Minister, and other political and business celebrities and foreign ambassadors, but the festivities were somewhat marred when local Liverpool MP, William Huskisson, was fatally injured after the *Rocket*, travelling on the westbound line, ran over his leg while his train had stopped at Parkside station. Huskisson died a few hours later in Eccles but the inaugural train continued on its way to Manchester where a large drunken crowd, angry at the death of the popular MP, pelted the Duke's coach with cabbages and potatoes.

Despite the opening day tragedy, as the world's first inter-city railway the Liverpool & Manchester Railway was an instant success, particularly with the travelling public. Goods services commenced three months later when Robert Stephenson's 2-2-0 inside-cylinder locomotive *Planet* was delivered from Newcastle – more powerful than the 0-2-2 *Rocket* it was capable of sustained speeds in excess of 35 mph. Within a short time the railway's success led to it carrying mail and Pickfords' inter-modal containers, as well as operating excursion trains and race-goers' specials. The opening of the railway brought another benefit to Manchester businesses as competition with the local canal companies – the Bridgewater Canal and the Mersey & Irwell Navigation in particular – forced them to cut their charges, which for some years before had been excessively high.

The Liverpool end of the line saw trains first using Crown Street station, reached initially through a single-line tunnel by means of a cable-hauled section, but this was replaced in 1836 by the more centrally located Lime Street station, also initially reached by a cable-operated incline. In Manchester the original terminus was at Liverpool Road station until the line was extended eastwards in 1844 to connect with the Manchester & Leeds Railway at Manchester Victoria station.

The success of the Liverpool & Manchester Railway (L&MR) soon led to proposals for a host of new railways throughout the land. Of these, the 82-mile Grand Junction Railway (GJR), from Birmingham to a junction with the L&MR at Newton, and the 112-mile London & Birmingham Railway were, by far, the most important. The former, engineered by George Stephenson and Robert Locke was authorized by Parliament in 1833 and opened on 4 July 1837. The latter, engineered by Robert Stephenson, was also authorized in 1833 but opened on 28 June 1838 after experiencing delays due to extensive engineering works at Tring Cutting (through the Chilterns) and Kilsby Tunnel (southeast of Rugby).

LEFT: Map of the Liverpool & Manchester Railway showing the route surveyed by George Stephenson.

BELOW: An early view of a train crossing Chat Moss on the Liverpool & Manchester Railway. The railway was built on a raft of wood and heather hurdles across the 12-square-mile peat bog.

The L&MR was absorbed by the Grand Junction in 1845. A year later the GJR merged with the London & Birmingham Railway forming the London & North Western Railway – the West Coast Main Line, as we know it today, was already taking shape.

Today the route of the L&MR is still in use as one of two rail links between those cities – the other is the former Cheshire Lines Committee route via Warrington, which opened in 1873. Recently electrified, the L&M route sees a regular timetable of passenger trains including the non-stop Transpennine Express.

BELOW: The original terminus at Liverpool was Crown Street station, which was reached via a rope-worked incline and a tunnel. It was replaced in 1836 by Lime Street station although Crown Street continued to be used for goods traffic until 1972.

RIGHT: Blasted out of sandstone rock, the 2-mile-long Olive Mount Cutting near Liverpool is up to 80 ft deep. The excavated rock was used to construct the Roby Embankment and the Sankey Viaduct.

Robert Stephenson and the London & Birmingham Railway

Engineer John Rennie proposed the first railway line between London and Birmingham as early as 1823. The route of the line would have been via Oxford and Banbury but neither this proposal nor a rival one via the Watford Gap and Coventry gained any financial backing. The two rival companies then joined forces and, in 1830, appointed Robert Stephenson as chief engineer. In order to avoid potential flooding along the Oxford route Stephenson chose the route through Coventry and the newly formed London & Birmingham Railway (L&BR) finally received Parliamentary approval in May 1833. The final route saw much modification due to pressure from canal companies and landowners – an example of this that can be seen today is Hemel Hempstead's station which is one mile from the town at Boxmoor, thereby avoiding crossing the land then owned by Sir Astley Cooper. In recent times the planned HS2 line between London and Birmingham has brought similar protests from people living along the proposed route. Nothing changes despite 180 years separating the two projects.

The railway's construction was a massive undertaking, involving 20,000 navvies for five years, a feat then calculated to exceed the effort that was required to build the Great Pyramid of Giza. Around one million cubic metres of earth and rock had to be excavated by men using only pickaxes, shovels and barrows in order to construct the 2½-mile-long Tring Cutting. To carry the railway over the valley of the Great Ouse River at Wolverton a massive 1½-mile earth embankment and viaduct were constructed. The building of the railway was a Herculean task involving the total excavation of 25 billion cubic feet of earth and rock.

The line was due to open at the same time as the Grand Junction Railway, which was due to reach Birmingham from the north, but completion of the L&BR was delayed by difficulties encountered in Northamptonshire with the construction of Kilsby Tunnel. A roof collapse and flooding caused by quicksand resulted in the 2,432-yd-long tunnel taking three years to build and ending up costing three times the original estimate.

Missing the coronation of Queen Victoria by nearly three months the railway finally opened throughout on 17 September 1838 with the first trains taking 5½ hours for the 112½-mile journey. Grand termini designed by Philip Hardwick were constructed at each end of the line: in London, Euston station was initially reached via a rope-worked incline powered by a stationary steam engine at Camden but this practice ended in 1844; in Birmingham, Curzon Street station was shared with the Grand Junction Railway (GJR), which provided services to and from Liverpool and Manchester. Locomotive workshops for the railway were provided at Wolverton.

Locomotives of the Lancashire builder Edward Bury initially worked the railway but this arrangement proved unsatisfactory and in 1839 Bury was appointed manager of the L&BR's Locomotive Department. His company in Liverpool also built forty-five locomotives for the new railway – these 0-4-0 tender locomotives featured a multi-tubed boiler and dome-topped firebox mounted on a lightweight wrought-iron inside bar frame, a design that became very popular with railroads in the USA.

By 1846 the L&BR had either taken on leases or bought outright several connecting railways and branch lines along its route. Probably the most important of these was the 51-mile Trent Valley Railway that linked Rugby with Stafford, allowing L&BR trains to bypass Birmingham, thereby offering a much faster service between London and Liverpool via the GJR. This route has stood the test of time and is today used by Pendolino trains which make the 193½-mile journey in only 134 minutes.

In that year the L&BR merged with the GJR and the Manchester & Birmingham Railway to form the London & North Western Railway, itself the largest company that was to become part of the newly formed London Midland & Scottish Railway in 1923. While today's railway travellers between London and Birmingham follow the original route of the L&BR in just 82 minutes, Hardwick's grand Euston station and Doric Arch were demolished amid massive protests in 1961–2. Fortunately Curzon Street station in Birmingham has survived and will be incorporated into the city's planned new HS2 station.

LEFT: Tring Cutting on the London & Birmingham Railway is 2½ miles long and an average of 40 ft deep. It was excavated by navvies using just picks, shovels and wheelbarrows with horses helping to pull men and barrows up the sides. The railway took 20,000 men five years to build before it was completed in 1838.

BELOW: A train on the newly opened railway leaves the grand castellated portico of Primrose Hill Tunnel in North London.

The Great Ventilating Shaft in Kilsby Tunnel in Northamptonshire, 1838. Eighteen working shafts were sunk to construct the 2,432-yd-long tunnel, which claimed the lives of twenty-six workers.

Robert Stephenson FRS

Born in Northumberland in 1803 Robert Stephenson was the only son of railway engineer George Stephenson, considered to be the 'Father of Railways'. He left school early to help his father resurvey the Stockton & Darlington Railway before attending Edinburgh University then working as a mining engineer in Colombia, South America. On his return to England Robert was appointed chief engineer at his father's steam locomotive building company in Newcastle, Robert Stephenson & Co. While his father was building the Liverpool & Manchester Railway Robert went on to develop the steam locomotive, *Rocket*, that went on to win the Rainhill Trials in 1829. In the same year he married Frances Sanderson but there were no children and she died from cancer in 1842. Following the success of the Rainhill Trials he was then appointed as chief engineer of the London & Birmingham Railway, an enormous engineering challenge that was completed in 1838.

In 1847 Robert Stephenson was elected as MP for Whitby, and remained in politics for the rest of his life. He was also a keen sailor and frequently travelled in his yacht, *Titania*, to Scandinavia and the Continent. By 1850 he had been involved in building around a third of Britain's railways, including the Chester & Holyhead Railway and the Newcastle & Berwick Railway, as well as designing the High Level Bridge at Newcastle, the Royal Border Bridge at Berwick-upon-Tweed and the wrought-iron Britannia Tubular Bridge across the Menai Strait in North Wales. He had also built railways in Belgium, Norway, France, Canada and Egypt before his untimely death in 1859. His funeral in London was a grand affair attended by thousands of mourners and he was buried in Westminster Abbey.

ABOVE: One of Edward Bury's 0-4-0 passenger steam locomotives is seen here at the London & Birmingham Railway's engine house in Camden Town in 1839. The roundhouse has survived and is now a performing arts and concert venue.

Joseph Locke and the Grand Junction Railway

Joseph Locke was born in Sheffield in 1805 and must rank as one of the most important railway engineers of the nineteenth century. His father, William Locke, was a notable mining engineer in northeast England and as a teenager Joseph gained much of his earlier skills working for him. In 1823 Joseph went to Newcastle to work for George and Robert Stephenson, who were then involved in the building of the Stockton & Darlington Railway. Even before this railway was completed George Stephenson had also been appointed surveyor and engineer of the newly formed Liverpool & Manchester Railway (L&MR) but his estimates of the cost of building the line were found to be seriously flawed and he was sacked.

The L&MR was resurveyed by Charles Vignoles while Joseph Locke was asked to resurvey the route of the tunnels under Liverpool. Both were highly critical of Stephenson's original survey but, despite this, George was eventually reappointed to build the railway. Both Vignoles and Locke became his assistants but the former soon resigned leaving the latter in charge of building the western portion of the line, which included the tunnels under Liverpool and the crossing of Chat Moss. On opening day in 1830 Joseph Locke was the driver of the *Rocket*, which fatally injured MP William Huskisson, while George drove the *Northumbrian*.

Grand Junction Railway

Joseph Locke, as assistant to George Stephenson, was appointed to survey the route of the proposed Grand Junction Railway (GJR) while the L&MR was being completed. Considered to be the world's first trunk railway, the new 82-mile GJR was to link Birmingham with the L&MR at Newton-le-Willows via Wolverhampton, Crewe and Warrington. It would meet up with the new London & Birmingham Railway (L&BR) at Birmingham, effectively creating a trunk railway linking Liverpool and Manchester with London. The gauge of all these railways was to be 4 ft 8½ in., the standard already set by George Stephenson.

The GJR received Parliamentary approval in 1833 and George Stephenson was appointed engineer for the project – while he took over responsibility for the southern half, his assistant Locke was to build the northern half. Unfortunately George's inherent weakness was his unprofessional approach to estimates and administration – in the past he had relied on his son Robert but by now the latter was carving his own solo career building the L&BR. On the other hand Locke was a meticulous planner and had prepared accurate estimates and issued contracts for his portion of line even before George had reached this stage. The GJR directors were not impressed and George resigned leaving Locke in charge of the whole project.

When it opened on 4 July 1837, The GJR was a showcase for Locke's administrative and engineering skills. The final

BELOW: Joseph Locke trained under George Stephenson before building a large number of railways in Britain and on mainland Europe.

RIGHT: Built by Joseph Locke from designs by George Stephenson, the Dutton Viaduct carried the Grand Junction Railway over the River Weaver in Cheshire.

figure for its construction came very close to his original estimate and it was built on time. There were few large engineering works required – the two largest being the 20-arch viaduct across the River Weaver and the 28-arch viaduct over the River Rea in Birmingham – with virtually no tunnels. Unlike George and Robert Stephenson he had a much greater faith in steam locomotives' abilities to haul loads up fairly steep gradients, in turn deeming fewer tunnels to be required. His later route of the Lancaster & Carlisle Railway over Shap Fell is testimony of this.

The GJR was completed before the London & Birmingham Railway, which had been delayed by problems encountered with the building of Kilsby Tunnel in Northamptonshire. When the latter opened in 17 September 1838, both companies used the new Curzon Street terminus station in Birmingham where adjacent platforms provided a quick interchange for passengers travelling from Manchester and Liverpool to Euston or vice versa.

In 1840 the GJR established its workshops at Crewe, then a small village set amidst Cheshire farmland, and in the same year took over the unfinished Chester & Crewe Railway. Paying handsome dividends to its shareholders the GJR went on to absorb the Liverpool & Manchester Railway and the North Union Railway from Newton-le-Willows to Preston, the latter a precursor to Joseph Locke's completion of the West Coast Main Line to Carlisle and Glasgow. In 1846 the GJR merged with the L&BR and the Manchester & Birmingham Railway to become the London & North Western Railway.

Joseph Locke's later life

Joseph Locke was in great demand by railway promoters across Britain following his completion of the Grand Junction Railway and went on to pursue a highly successful career building railways both at home and overseas. As the engineer for the Lancaster & Preston Junction Railway, the Lancaster & Carlisle Railway and the Caledonian Railway (between Carlisle and Glasgow/Edinburgh) his greatest triumph was undoubtedly the completion of the West Coast Main Line north of Birmingham – in line with his belief in the power of steam locomotives to tackle steep gradients, the whole route was achieved without the building of one tunnel, quite a remarkable achievement.

Locke was also appointed to build the Manchester & Sheffield Railway through the 3-mile Woodhead Tunnel, the London & Southampton Railway and the Glasgow, Paisley & Greenock Railway, while on the Continent he planned and built railways in France, Spain and Holland. He also dabbled in politics, being elected as MP for Honiton in Devon in 1847. Following an illustrious career as one of Britain's foremost railway engineers, Locke died in 1860 and is buried in Kensal Green Cemetery in West London.

RIGHT: This lithograph drawn on stone by E Bissell depicts a train calling at Newton Road station, West Bromwich, near Birmingham, on the Grand Junction Railway shortly after opening in 1837.

Edinburgh & Glasgow Railway

The first proposal for a railway between Scotland's two largest cities, Glasgow and Edinburgh, came from the engineer Robert Stevenson as early as 1817. While better known for his lighthouses that were built around the coast of Scotland, Stevenson also went on to explore a number of possible routes for canals or railways between Stirling and Aberdeen in eastern Scotland. While nothing came of Stevenson's proposal for the Glasgow to Edinburgh route, the first detailed survey for an intercity railway was undertaken by James Jardine in 1826 – Jardine went on to build the Innocent Railway in Edinburgh, which opened in 1831 and featured Scotland's first railway tunnel.

In 1830 a further survey was carried out by the engineering partnership of Thomas Grainger and John Miller but the proposal to build the railway was subsequently rejected by Parliament. The project was then shelved until 1838 when John Miller, then aged 33, was appointed as engineer and an amended route presented before Parliament. The Edinburgh & Glasgow Railway Company was this time successful in its bid to build the line which received Parliamentary approval in July of that year.

Apart from its western approach into Glasgow, Miller's final route between Haymarket station in Edinburgh and Queen Street station in Glasgow via Falkirk was consistently level and featured seven viaducts and three tunnels. One such tunnel was built near Falkirk purely to hide the railway from the view of a local landowner, a practice that was soon to be repeated many times across Britain, resulting in increased construction costs. The largest engineering feature on the line is the curving Ratho Viaduct that carries the railway over the Almond Valley between Broxburn and Newbridge. A central embankment separates the thirty-six arches on the eastern side and the seven arches on the western side. At the Glasgow end of the line the railway descends down the 1-in-44 Cowlairs Incline and through a 1,040-yd tunnel to terminate at Queen Street High Level station. The locomotive, carriage and wagon works for the railway was established at Cowlairs in 1841.

The Edinburgh & Glasgow Railway (E&GR) opened to the public on 21 February 1842 and was an immediate success. In Edinburgh it was extended eastwards from Haymarket to link up with the North British Railway at what is now known as Waverley station in 1846. However, the steep Cowlairs Incline and tunnel that trains had to negotiate on leaving the cramped confines of Glasgow's Queen Street station remained a weak link in the chain for 120 years. Edinburgh-bound trains were hauled up through the tunnel and incline by ropes attached to a stationary steam winding engine at Cowlairs, apart from a short and unsuccessful experiment with banking engines in the late 1840s. In 1909 cable haulage ended with the introduction of more powerful steam banking engines. These remained in service until more powerful diesel locomotives capable of hauling their trains unassisted were introduced in the 1960s.

The E&GR went on to take over several railways in Central Scotland and in 1864 was absorbed by the North British Railway. Together with his business partner Thomas Grainger the railway's builder, John Miller, went on to construct many railways around Scotland including those from Glasgow to Dumfries, Dundee to Arbroath and Edinburgh to Berwick upon Tweed. He was one of two MPs for Edinburgh between 1868 and 1874 and died in 1883. Today his 47¼-mile line between Edinburgh and Glasgow is by far the busiest and quickest of the three routes linking the two cities and electrification is planned for completion by 2017.

TOP RIGHT: After opening to Edinburgh Haymarket station in 1842, the Edinburgh & Glasgow Railway was extended eastwards in 1846 to link up with the North British Railway at what is now known as Waverley station.

BOTTOM RIGHT: The major engineering feature on the Edinburgh & Glasgow Railway is the graceful 43-arch Ratho Viaduct across the Almond Valley between Broxburn and Newbridge. The railway is today one of three routes that link Scotland's two largest cities although a fourth, high-speed, line is proposed for the future.

Newcastle & Carlisle Railway

The 63-mile Newcastle & Carlisle Railway was one of the earliest public railways in Britain, receiving Parliamentary approval in 1829. A proposal to build a canal linking the two cities had been made at the end of the eighteenth century but the railway was seen as a much cheaper option. However, the new railway had one serious drawback – in order to appease the wealthy landowners along its route the 1829 Act of Parliament forbade the use of steam locomotives.

The railway was built in stages, with the 16¾ miles between Blaydon and Hexham opening to passengers on 9 March 1835. Despite steam locomotives being forbidden the company had ordered three and the sight of these snorting monsters so enraged a local landowner that he applied for an injunction to stop their use. Much to the local population's chagrin and less than three weeks after its opening, the Blaydon to Hexham train service ground to a halt. The railway was successful in persuading Parliament to amend the original Act allowing it to use steam locomotives and services restarted on 6 May.

Construction work continued along the rest of the mainly double-track route and was completed in 1838. The official opening day on 18 June was a grand affair with no less than 3,500 passengers being carried in thirteen trains between the railway's eastern terminus at Redheugh, south of the River Tyne in Newcastle, and London Road station in Carlisle. Accommodated in open wagons fitted with benches, they had a pretty miserable trip in the persistent rain! A year later a short branch was opened from Blaydon to a temporary station at Newcastle Shot Tower via a timber bridge over the Tyne at Scotswood.

The British class system prevailed even in those very early days of rail travel. Those able to afford first-class fares enjoyed covered carriages with doors while second-class passengers just had a roof over their heads. Poor third-class passengers had to sit in open goods wagons fitted with benches, open to all weathers and the smoke and hot cinders from the locomotive. Issuing tickets on hand-written pieces of paper was a laborious job so even before the line was completed the station master at Milton, a certain Thomas Edmondson, had introduced printed and numbered card tickets – not only saving booking office procedures but also creating an accurate record of fares collected, a system which remained in use for 150 years.

The N&CR was a great success and paid a handsome dividend to its shareholders during its lifetime, the

directors being very much aware of the tourism opportunities that it offered on its scenic route paralleling Hadrian's Wall. A 105-page Companion was published a year before the railway was completed describing the route, with short guides to the cities at each end and details of inns near the intermediate stations. With a journey time of three hours the railway not only offered six return passenger trains on weekdays but also two on Sundays, the latter being operated despite the wrath of the newly founded Lord's Day Observance Society!

Temporary arrangements at Newcastle were eventually replaced at the beginning of 1851 when N&CR trains started using the newly opened Central station – this grand, classical style station, built jointly by the York & Newcastle Railway and the Newcastle & Berwick Railway, had been opened a few months before by Queen Victoria. At Milton the N&CR formed a junction with the Earl of Carlisle's Waggonway, a 10-mile horse-drawn coal-carrying wooden wagonway that had opened as early as 1775. It was re-laid with iron rails in 1809 and a horse-drawn passenger service was introduced between Milton (later renamed Brampton Junction) and Brampton Town in 1836, with steam haulage on the 1-mile branch line being introduced in 1881. At Carlisle the N&CR was extended westwards from London Road station to the Canal Basin in 1837, allowing transshipment of coal on to barges. The N&CR also went on to open a 13½-mile branch line from Haltwhistle to Alston in 1852.

The N&CR remained independent until 1862 when it was absorbed by the North Eastern Railway (NER). Hexham became an important junction with the opening throughout of the North British Railway's meandering cross-border line to Reedsmouth and Riccarton Junction in the same year, and the Hexham & Allendale Railway's 12¼-mile branch to Allendale in 1869. The Earl of Carlisle's Brampton Town branch line was taken over by the NER in 1912 but had closed to passengers in 1923. The rest of the colliery line remained open until it was closed by the National Coal Board in 1953.

Since it was first opened in 1839 the bridge over the River Tyne at Scotswood has had a chequered history – the first wooden bridge was destroyed by fire in 1860, its (wooden) replacement lasting until 1865 when a temporary single-track bridge was opened. This in turn was replaced by a 6-span wrought-iron bridge in 1871 which closed permanently in 1982 when trains were diverted via Dunston-on-Tyne and over the King Edward VII Bridge to Central station. The rusting bridge at Scotswood still stands today.

Connecting the East Coast and West Coast Main Lines the railway between Newcastle and Carlisle and the fourteen intermediate stations are still open for business. Many of the stations still feature original NER ornate lattice-metal footbridges, Tudor-style station buildings and distinctive wooden signal boxes. The double-track line is also a useful diversionary route during closure of either of the two Anglo-Scottish main lines.

FAR LEFT: A busy scene at Hexham goods depot on the Newcastle & Carlisle Railway in 1836. The 63-mile line was completed in 1838 and is still open for business today.

LEFT: A poster advertising cheap return fares on the Newcastle & Carlisle Railway during the Whitsuntide Holiday in May 1847.

OVERLEAF: With views across the River Tyne to an ironworks, a steam locomotive on the Newcastle & Carlisle Railway in 1836 hauls a fully loaded train at Wylam Scars near Newcastle.

Great Western Railway

The port of Bristol was a busy place by the early nineteenth century with ships plying their international trade across the Atlantic and beyond. Communications between the city and other parts of Britain, in particular London, were poor apart from the slow and often dangerous seagoing route around Lands End. The uncomfortable stagecoach journey along turnpike roads could take at least a day even by the late eighteenth century, when the first mail coach was introduced. As early as 1824 a scheme to link the two cities by railway had been proposed but this came to nothing, as did another in 1832. The turning point came in March 1833 when a group of Bristol businessmen appointed a young engineer, 27-year-old Isambard Kingdom Brunel, to survey the route of yet another proposed railway to London. By August the survey of a near-level route via Bath, the village of Swindon and the Thames Valley to Paddington had been completed, two boards of directors established (one at Bristol and one at Paddington) and the Great Western Railway (GWR) was born.

Parliament finally authorized the building of the London to Bristol line along Brunel's surveyed route on 31 August 1835 after an initial rejection of the scheme. No gauge had been stipulated in the Act and two months later the GWR directors opted to build their line for speed and comfort to the broad gauge of 7 ft 0¼ in. Unfortunately, apart from in the south west, most other railways around Britain were built to what became the standard gauge of 4 ft 8½ in. Brunel appointed Daniel Gooch as the GWR's first chief mechanical engineer in 1837, a position to be later held by such luminaries as William Dean, George Jackson Churchward, Charles Collett and Frederick Hawksworth.

The 118¼-mile railway was built in nine sections with the first from Paddington to Maidenhead opening on 4 June 1838, to Twyford on 1 July 1839, to Reading on 30 March 1840, to Steventon on 1 June, to Faringdon Road on 20 July, to Hay Lane (near Swindon) on 17 December and to Chippenham on 31 May 1841. From the Bristol end the section to Bath from Temple Meads opened on 31 August 1840 but the complete opening of the line was delayed by the construction of Box Tunnel, not opening until 30 June 1841. 4,000 navvies were working on the 3,212-yd-long tunnel at the peak of its construction, of whom around 100 died in accidents. It was also built with great accuracy and when the two ends met underground they were less than two inches adrift. It is said that Brunel deliberately aligned the tunnel so that the sunrise could be seen through it when viewed from the western end at Box on his birthday, 9 April. The other major engineering feat on the line was Sonning Cutting, east of Reading. The cutting is over a mile long and up to 60 ft deep and was excavated by hundreds of navvies using just picks, spades and wheelbarrows, taking them two years to complete it.

Brunel designed imposing termini fit for his modern railway at each end of the line. His train shed at Bristol

LEFT: Appointed as Chief Engineer of the Great Western Railway at the age of 27, Isambard Kingdom Brunel was one of Victorian Britain's foremost engineers.

TOP RIGHT: Designed by Brunel, the masonry arched bridge over the River Thames at Maidenhead has stood the test of time and today carries high-speed trains between Paddington and Bristol.

BOTTOM RIGHT: A view of Brunel's timber-roofed station at Temple Meads in Bristol in 1846. Although not now used by trains, the Grade I structure still stands and is today used as a covered car park for the adjacent former GWR station.

Temple Meads still stands, albeit as a car park and without any train tracks. Much of Paddington station as well know it today was also designed by Brunel along with his associate Matthew Wyatt but was not completed until 1854. There were no stations at Swindon or Didcot when the line opened but this was all soon to change when the former became the junction for what became the GWR's route to South Wales via Gloucester and the location of the company's main works. Didcot soon developed as an important junction for the Oxford, Worcester and Birmingham lines. Swindon grew from a small village, to become the largest railway town in Britain, with the railway works in its heyday in the 1930s employing around 14,000 people.

The first locomotives built for the GWR were a motley collection of 2-2-2s ordered by Brunel from six different British manufacturers. On his appointment as chief mechanical engineer in 1837 Daniel Gooch initially purchased twelve 'Star' Class 2-2-2s from Robert Stephenson & Co. before he designed several classes of locomotives using standardized parts such as the highly successful 'Firefly' and 'Sun' passenger classes – the former was capable of the then unheard of speed of 60 mph. The opening of Swindon railway works in 1843 soon ushered in a golden age of steam locomotive building that finally ended in 1960 when British Railways' last steam locomotive, appropriately named *Evening Star*, was rolled out.

It became obvious by the 1860s that Brunel's unique broad gauge was now out of step with the rest of the nation's expanding standard gauge system – the 4-ft 8½-in. gauge had already been made mandatory for new railways in Great Britain in the Regulating the Gauge of Railways Act 1846. The problem was eventually addressed after Brunel's death by gradually converting GWR track to standard gauge between 1864 and 1892. The GWR prospered and within a few years had spread its tentacles across southwest England, the West Midlands and South Wales.

Brunel's Paddington to Bristol route has stood the test of time and it was announced in 2012 that the line is to be electrified. It will be one of the last major electrification schemes on Britain's railways, due to be completed in 2017 when the aging InterCity 125 diesel train sets, first introduced over forty years ago, will be replaced by the new Hitachi Super Express electro-diesel multiple units.

BELOW: The *Tartar*, a 4-2-2 broad-gauge locomotive of the 'Iron Duke' Class was designed by Daniel Gooch and built at the Great Western Railway's locomotive works at Swindon in 1848.

RIGHT: Built for the GWR in 1842 by Fenton, Murray & Jackson, the broad-gauge 2-2-2 locomotive *Acheron* of the 'Firefly' Class emerges from Box Tunnel on the mainline between Bristol and London.

THE
GREAT WESTERN RAILWAY
BY
J.C.BOURNE

Manchester & Leeds Railway

Transporting goods over the Pennines between Lancashire and Yorkshire had been a time-consuming business until the late 1700s, the centuries-old rutted tracks impeding progress towards the Industrial Revolution. In 1776 this all changed when a group of Rochdale businessmen proposed the building of a canal from Manchester to Sowerby Bridge, where it would meet the Calder & Hebble Navigation which had opened in 1770. The 32-mile Rochdale Canal and its ninety-two locks was engineered by James Brindley and opened throughout in 1800, linking up with the Bridgewater Canal at Manchester. Being built as a wide canal it had serious advantages over its rival, the Huddersfield Narrow Canal, which burrowed under the Pennines through the 5,696-yd-long Standedge Tunnel.

The Rochdale Canal was a success from the start and, until the coming of the railways, became the premier route for trade between Lancashire and Yorkshire, carrying vast amounts of coal, wool, cotton, timber, salt and limestone along with finished goods from the hundreds of textile mills – it carried just under one million tons at its peak in 1845.

The canal's supremacy came to an end with the opening of the Liverpool & Manchester Railway (L&MR) in 1830. Vast quantities of raw materials and finished goods could be transported quickly and cheaply by this revolutionary new transport system between the port of Liverpool and east Lancashire. Soon further schemes were put forward to build railways from Manchester to the textile mills in the Pennine foothills and further afield across the Pennines to Yorkshire. Three major railway routes were built in the end, all tunnelling through the hills, but the first to see the light of day was the Manchester & Leeds Railway (M&LR), which in 1836 received Parliamentary authorization.

Engineered by George Stephenson and Thomas Gooch (brother of GWR locomotive superintendent Daniel Gooch), the railway was opened in stages: from Oldham Road station in Manchester to Littleborough in 1839; from Normanton (where it met the North Midland Railway's line to Leeds) along the Calder Valley to Hebden Bridge in 1840; and the intervening gap between Littleborough and Hebden Bridge, which involved the excavation of Summit Tunnel, in 1841. In 1844 the line

was extended at its western end to meet the L&MR's new station at Hunt's Bank (later renamed Manchester Victoria). Travellers between Manchester and Leeds had to suffer a 6½-hour journey by horse-drawn coach before the coming of the railway – the new railway journey was slashed to 2¾ hours but was more expensive. Running trains on Sundays also caused problems, with the railway's chairman and two directors resigning in protest at this desecration of the Sabbath. Sadly the railway soon had a poor reputation for its treatment of passengers – 'herded like pigs and sheep into open third-class carriages' – and its lack of punctuality.

The M&LR became the chief constituent company of the Lancashire & Yorkshire Railway (L&YR) in 1847 after absorbing six railways in the region. In 1855 the wheel turned full circle when the L&YR, along with three other railway companies, took on a 21-year lease of the Rochdale Canal.

Summit Tunnel was for a short period the longest railway tunnel in the world, at just over 1½ miles long – three months later it was eclipsed by the opening of Brunel's Box Tunnel on the Great Western Railway. Excavated through treacherous seams of shale, sandstone and coal by a team of navvies using just picks, shovels and wheelbarrows, Summit Tunnel was lined with over twenty-three million bricks and took three years to complete. Nine navvies lost their lives during its construction. The tunnel has only been closed once since opening in 1841, when on 20 December 1984 a derailed petrol train caught fire inside it. Served by diesel trains on the Calder Valley Line, the Manchester to Leeds route is alive and well today.

LEFT: Engineered by George Stephenson, the Manchester & Leeds Railway opened in 1839. The bridge depicted in this lithograph carries the railway over the River Irwell in Manchester.

BELOW: The construction of the Manchester & Leeds Railway involved the building of many bridges, viaducts and tunnels. Depicted in this lithograph the railway crosses the Rochdale Canal on the 9-arch Todmorden Viaduct.

Early railways in Southern England

Canterbury & Whitstable Railway

The city of Canterbury in Kent for centuries relied upon the River Stour for its supply of goods and raw materials. Horse-drawn barges often took several days to make the meandering 70-mile journey from Ramsgate along a continuously silting-up river. The state of the river had reached crisis point by the early nineteenth century but early railway pioneer William James made a proposal to build a line from the city to the port of Whitstable, only eleven miles to the north. During 1823 and 1824 he surveyed three possible routes but bankruptcy and illness prevented him from any future involvement. George Stephenson was duly appointed engineer of the newly formed Canterbury & Whitstable Railway (C&WR), which received Parliamentary approval in 1825.

Stephenson, assisted by Joseph Locke, opted to build the most direct route surveyed by James but this included steep gradients incapable at that time of being worked by steam locomotives. Stationary steam engines were planned to work the three inclines with horsepower on the level sections. In 1828 construction work commenced and the standard-gauge line was opened for passengers and goods on 3 May 1830, just over four months before the opening of the Liverpool & Manchester Railway. A decision was taken to use a steam locomotive on the incline out of Whitstable Harbour and *Invicta* was delivered by ship to Whitstable from Stephenson's Newcastle works and headed the inaugural train; however, it proved incapable of hauling loads on this northerly section of the line and, despite modifications, was soon replaced by a stationary steam engine.

Facing bankruptcy the C&WR was taken over by the newly formed South Eastern Railway in 1844. Two years later the line was rebuilt to allow the use of steam locomotives along its entire length, although only specially adapted locomotives could be used due to the limited height of the Tyler Hill tunnel.

London & Greenwich Railway

Receiving Parliamentary approval in 1833 the 3¾-mile London & Greenwich Railway (L&GR) opened in stages between 1836 and 1838. Its elevated route ran from Tooley Street terminus near London Bridge to Deptford and Greenwich, crossing Deptford Creek by a manually operated lifting bridge. Designed primarily for passengers the L&GR was not only the first to use steam locomotives in London but was also the first elevated railway in the world, its entire length being supported on 878 brick arches. From 1845 until 1923 the L&GR was leased by the South Eastern Railway. Tooley Street terminus was replaced by nearby London Bridge station, which was opened by the London & Croydon Railway in 1839 – today London Bridge handles over fifty-four million passengers a year, the fourth busiest station in the UK.

London & Southampton Railway

Originally conceived as a canal, the London & Southampton Railway emerged in 1831 as the Southampton, London & Branch Railway & Dock Company. In 1834 construction of the line between Nine Elms and Southampton commenced under the direction of the company's first engineer, Francis Giles. Progress was painfully slow, however, and as costs escalated the company became a laughing stock. Giles resigned in 1837 and was replaced by Joseph Locke, already renowned for his work on the Stockton & Darlington and Liverpool & Manchester railways.

The line opened in stages, the first section from Nine Elms to Woking opening on 21 May 1838. It was extended to Winchfield on 24 September and to Basingstoke on 10 June 1839, the same day that the Southampton to Winchester section also opened. Amidst great rejoicing along the line, the intervening gap between Winchester and Basingstoke opened on 11 May 1840.

Meanwhile in 1839 the London & Southampton Railway had changed its name to the London & South Western Railway when it was authorized to build a railway from Bishopstoke (later Eastleigh) to Portsmouth. The original London terminus at Nine Elms, designed by Sir William Tite who also designed the Southampton terminus, soon proved to be insufficient for the rapidly expanding railway and, in 1845, the LSWR received authority to extend the line to a new terminus at Waterloo. This opened in 1848, the Nine Elms site being expanded to include a goods depot and the company's locomotive works which remained there until 1910 upon transferal to a new site at Eastleigh.

With third-rail electrification completed in 1967, the Waterloo to Southampton mainline is today a busy commuter and freight route with the London terminus, the busiest in the UK, handling ninety-four million passengers a year.

ABOVE: Seen here crossing Corbett's Lane, London's first railway, the London & Greenwich, opened in stages between 1836 and 1838. Designed as a commuter railway, its entire length was supported on 878 brick arches.

LEFT: Robert Stephenson's locomotive *Invicta* hauls the first train on the Canterbury & Whitstable Railway in Kent in 1830. Engineered by George Stephenson, the railway opened four months before the Liverpool & Manchester Railway.

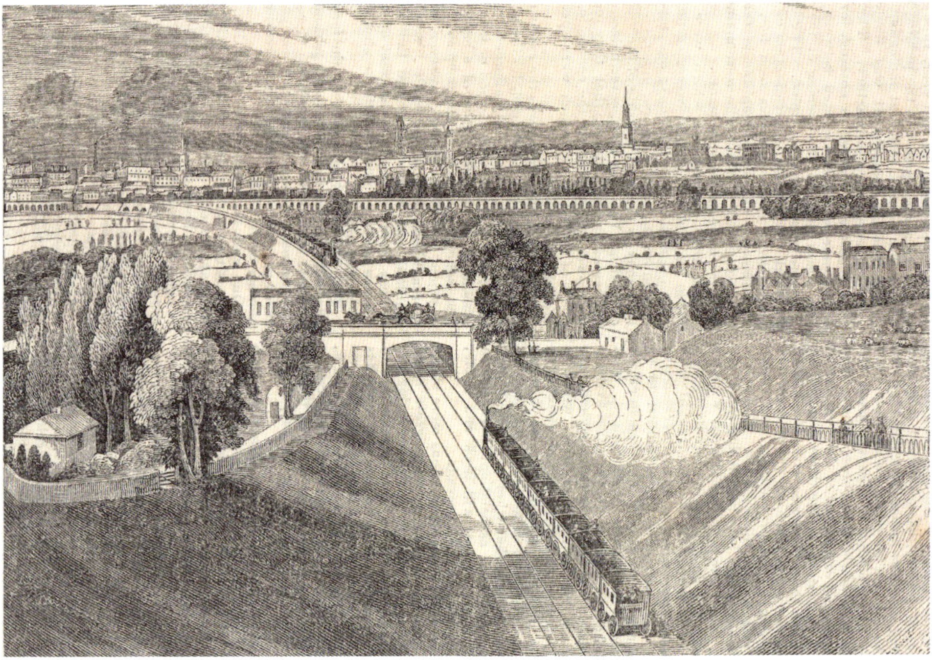

London & Croydon Railway

Receiving Parliamentary approval in 1836, the 8¾-mile London & Croydon Railway (L&CR) followed the course of the disused 1809 Croydon Canal. The railway company appointed William Cubitt as engineer but the line cost 340 per cent more than the original estimate of £180,000 due to the construction of long cuttings at Forest Hill and New Cross. The 4-ft 8½-in.-gauge railway between Croydon and Corbetts Lane junction with the London & Greenwich Railway (L&GR) via New Cross opened in 1839. The railway had running powers over the L&GR in order to reach the planned terminus at London Bridge. At New Cross the L&CR established an engine shed and a freight depot, located alongside the Surrey Canal.

The L&CR soon became an important funnel for two other railways serving London Bridge. Joining it at Norwood, the London & Brighton Railway (L&BR) opened throughout in 1841 while the South Eastern Railway's (SER) route from Dover, which also joined the L&CR at Norwood, was completed in 1844.

Concerned about the high charges extracted from them by the L&GR for access to London Bridge, the SER and the L&CR joined forces in 1843 to build a short branch line from New Cross to a new terminus at Bricklayers' Arms. The imposing new station designed by Lewis Cubitt opened in 1844 but a year later the L&GR reduced its charges and the L&CR started using London Bridge again. The SER continued to use Bricklayers' Arms terminus until 1852 by which time it had already absorbed the L&GR. The SER's large 1844-built engine shed and repair shop continued in use until closure by British Railways in 1962.

In 1846 the L&CR, the L&BR along with two other companies amalgamated to form the London, Brighton & South Coast Railway (LB&SCR).

ABOVE: Seen here near New Cross, the London & Croydon Railway was engineered by William Cubitt and opened in 1839 between Croydon and Corbett's Lane – at the latter it formed a junction with the London & Greenwich Railway allowing trains from Croydon to operate to London Bridge station.

RIGHT: The ornate Gothic-style pumping houses on the London & Croydon's atmospheric railway had a short working life of only two years before the system was abandoned in 1847.

London & Croydon's atmospheric railway

The L&CR briefly experimented with an atmospheric railway, laid alongside its increasingly congested line between Dartmouth Arms (Forest Hill) and West Croydon. The first railway flyover in the world was built to carry the atmospheric line over the existing steam line at Norwood. Dispensing with locomotives, the system entailed a slotted steel pipe in the middle of the track – inside the pipe was a piston, connected to the underside of the leading coach by an iron rod, which was sucked along by a permanent vacuum created by three steam-driven Gothic-style pumping houses set at intervals along the route. In theory the slot in the top of the pipe was sealed with air-tight leather flaps, which first opened and then closed as the train moved forward.

Although initial testing in November 1845 was successful the system soon proved to be a resounding failure after two pumping engines failed. The leather flaps were also highly ineffective at ensuring a permanent air-tight fit – not only did rats find the leather very tasty, the flaps dried and cracked during warmer weather. In 1846 its new owner, the newly formed LB&SCR, cancelled a planned extension to Epsom but then curiously extended the system from Dartmouth Arms to New Cross in January 1847. By now the writing was on the wall and in May of that year the system was finally abandoned. Down in Devon Brunel's atmospheric South Devon Railway was still being extended but its days were also numbered.

1843

GWR's locomotive works at Swindon opens

1844

Bricklayers' Arms terminus in south London opens

Bristol & Exeter Railway opens

GWR Swindon Mechanics Institute founded

Midland Railway formed

Yarmouth & Norwich Railway opens

1845

Northern & Eastern Railway extended to Cambridge and Brandon

London & North Western Railway formed

Peak of 'Railway Mania'

1846

Regulating the Gauge of Railways Act

EXPANSION and 'RAILWAY MANIA'

1843–1850

1847
- Atmospheric South Devon Railway (SDR) opens
- Greenwich Mean Time used to standardize railway timetables
- Lancashire & Yorkshire Railway formed

1848
- George Stephenson dies
- West Coast Main Line completed between London and Glasgow

1849
- Eastern Counties Railway opens Norwich Victoria station

1850
- Chester & Holyhead Railway completed
- Glasgow & South Western Railway (G&SWR) formed

George Hudson and 'Railway Mania'

Following the financial ruin that had befallen speculating investors in the South Sea Company, otherwise known as the 'South Sea Bubble', the British Government introduced the Royal Exchange and London Assurance Corporation Act (commonly known as the 'Bubble Act') in 1719. Aimed at preventing similar speculation and fraudulent trading in worthless business ventures, the Act was designed to limit their formation and required all new joint-stock companies to be incorporated by Act of Parliament. The 'Bubble Act' was repealed by Parliament in its wisdom in 1825.

Following the opening of the Liverpool & Manchester Railway in 1830 and its subsequent success, proposals for other such railways around Britain soon followed. However, brakes were put on railway development due to banks not lending money in times of political and social unrest, and a sluggish British economy where the high interest rates offered by Government bonds were more attractive investments.

By the early 1840s the British economy was bouncing back, interest rates were cut and the manufacturing industry was booming – the period of the Victorian entrepreneur had dawned. Investors who had previously put their money into Government bonds were soon attracted by the success of the few already established railways, such as the London & Birmingham, the Grand Junction and the Great Western.

The rapid rise of railway shares soon saw a speculative frenzy in a plethora of overambitious and sometimes fraudulent proposals for new railways across Britain. Known as 'Railway Mania' it reached its peak when 272 Acts of Parliament were passed in 1846, authorizing 9,500 miles of new railway – in the end just one third were built due to the premature collapse of railway companies. This level of authorization was unsurprising with so little Government regulation, as many MPs such as the 'Railway King' George Hudson were directors of these fledgling railways. The general public also saw a potentially quick profit, often using life savings to purchase shares for only a 10 per cent deposit while the railway companies had the right to call in the remainder when they pleased. Many prospective schemes collapsed, sadly leaving the investor to stump up the balance of an already worthless investment – the bubble had burst and, not for the last time, the British middle classes had become the victims, as greedy investors would find out to their cost in the late twentieth-century 'Dot-Com Bubble Crash'.

By the early 1850s new railway construction was chiefly limited to established railway companies such as the Great Western Railway, the Midland Railway and the London & North Western Railway who had profited by buying failed companies for a fraction of their original value. On a more positive note many of those railways actually formed the basis of today's railway network, leaving a tangible and strategically important legacy despite the boom-and-bust of the 'Railway Mania' period.

A Director of the Diddle-cum-profit Railway.

PREVIOUS SPREAD: View of the Midland Railway, 1845 – steam locomotives hauling wagons marked 'Johnson Cammell & Co' and a locomotive hauling a passenger train are shown in front of the massive Cyclops Steel Works in Sheffield. Formed in 1844 as an amalgamation of three railway companies, the Midland Railway's driving force was George Hudson, the 'Railway King'.

LEFT: 'A Director of the Diddle-cum-profit Railway' – one of several satirical sketches used for headed notepaper satirizing the speculative investment fever of the railway mania that took place between 1845 and 1846 in which many investors lost their life savings.

George Hudson

George Hudson, a farmer's son, was born in 1800 and by the age of 27 had become a director in a firm of drapers in York, one of the city's largest employers. In 1827 he inherited a fortune from his great uncle and went on to establish the York Union Banking Company. In 1835 Hudson was elected to York City Council, becoming Lord Mayor two years later.

After forming a business relationship with George Stephenson, Hudson promoted the York & North Midland Railway which was authorized in 1837. He became the largest company shareholder and was instrumental in persuading Stephenson to route the line through York instead of bypassing it. He was also a major shareholder in three other railways – the Midland Counties Railway, the North Midland Railway and the Birmingham & Derby Junction Railway – forming in 1844 the Midland Railway, also led by Hudson. Hudson's York & Newcastle Railway – today's East Coast Main Line – was also nearing completion north of York. By 1846, at the peak of 'Railway Mania', he controlled a 1,000-mile railway empire and was known widely as the 'Railway King'. His wealth allowed him to indulge in Yorkshire's 12,000-acre Londesborough Estate. In 1847 he opened his own private station serving the estate when the York & North Midland Railway opened between York and Market Weighton. Not content with his railway empire-building, and by now a personal friend of the Duke of Wellington, Hudson was also elected MP for Sunderland in 1845, a position he held until 1859 when he eventually fell from power.

Hudson tabled new railway proposals for thirty-two Parliamentary Bills a year after his election to Parliament – soon his companies controlled at least 25 per cent of English railways. However, there was large-scale fraud and corruption behind this success. After the 'Railway Mania' bubble burst at the end of 1847 Hudson's world fell apart when he was forced to retire as chairman of the railway companies he owned, being forced to sell his Londesborough Estate two years later. A committee of investigation found that the unscrupulous Hudson had bribed MPs to vote for his railway projects, manipulated share prices by using insider information, sold land he didn't own and paid dividends from capital. York City Council expelled him in disgrace – even his waxwork lookalike at Madame Tussauds was destroyed. Imprisoned for debt in York Castle in 1865 after admitting his guilt, Hudson was released after only three months when his few remaining friends raised the necessary funds. In 1871 he died a broken man. However, his legacy was to establish York as a major railway centre, a position it still holds today.

ABOVE: By wheeling and dealing – often illegally – George Hudson ended up controlling an enormous railway empire that stretched from Birmingham to northeast England. Known as the 'Railway King', his fall from grace in 1847 was sudden and he ended his years living off hand-outs from the few friends that he had left in the world.

The Midland Railway

Formed in 1844, the Midland Railway (MR) went on to become the third largest pre-grouping railway company in the UK with its tentacles reaching as far south as Bournemouth and London, eastwards into Norfolk, westwards into the South Wales Valleys, northwards to Yorkshire and Scotland and even into Northern Ireland. It also became the country's most important coal carrier, with shrewd takeovers and the opening of new lines giving access to the vast Midlands' coalfields.

The MR had its origins in three early railways. The first to be built was the Midland Counties Railway (MCR), which had its origins in the horse-drawn coal-carrying Mansfield & Pinxton Railway (M&PR) that had already opened in 1819. The M&PR proposed a new railway that would link Nottingham, Leicester and Derby, seeking further outlets for locally mined coal. A southward extension from Leicester to Rugby where it would meet the London & Birmingham Railway, then under construction, was added to the proposal that eventually received Parliamentary approval in 1836 after several failed attempts. The MCR opened between Derby and Nottingham in 1839 and Trent Junction to Rugby in 1840. A year later Thomas Cook organized his first ever rail excursion using the MCR to carry over 500 members of a temperance movement from Leicester to a rally in Loughborough.

The second of the MR's founding railways was the Birmingham & Derby Junction Railway (B&DJR). The route between Derby, Burton-upon-Trent, Tamworth, Whitacre and Hampton-in-Arden, where it would meet the London & Birmingham Railway (L&BR), surveyed by George Stephenson in 1835, received Parliamentary approval in 1836. This was undoubtedly helped by the fact that the MP for Tamworth, Robert Peel, was also Prime Minister! From Hampton-in-Arden B&DJR trains were to reverse direction to reach Curzon Street station in Birmingham, then jointly owned by the L&BR and the Grand Junction Railway. The railway opened in 1839, offering a service between Derby and London which, within a year, was in direct competition with the newly opened Midland Counties Railway's route via Leicester and Rugby. A price war soon erupted between the two companies for passenger and coal traffic to London but, in the end, neither side emerged as the winner. The railway stopped using the joint Curzon Street station in 1842 when it opened its own

terminus at Lawley Street – a Freightliner terminal today stands on the site.

The third of the MR's founding railways to be built was the 72-mile North Midland Railway (NMR) between Derby and Leeds, which was also surveyed by George Stephenson in 1835 and authorized by Parliament a year later. Following major river valleys for much of its route it involved the building of seven tunnels, 200 bridges and a canal aqueduct, meeting the Sheffield & Rotherham Railway at Masborough and George Hudson's York & North Midland railway at Leeds–with its heavy engineering and miles of embankments and cuttings, the railway was expensive to build.

Opening in 1840, the railway soon ran into financial difficulties and by 1842 a group of shareholders, most prominent among them being George Hudson the 'Railway King', had forced severe economies including the closing of stations, a reduction in services and the sacking of many key staff. Hudson had effectively taken over control of the NMR.

George Hudson lost no time in expanding his empire and came to a behind-the-scenes agreement with the B&DJR to take it over. The MCR was now in a no-win situation and with Hudson pulling the strings the company's shareholders caved in and voted for an amalgamation with the two other railways. Thus was the Midland Railway born. The story of the company's later expansion northwards with its own Anglo-Scottish route via the Settle–Carlisle Line is told in the next chapter.

One casualty of the formation of the MR in 1844 was the B&DJR's short section of line between Whitacre Junction, near Birmingham, and Hampton-in-Arden. With MR traffic between Derby and London now taking the former MCR route via Rugby, the line to Hampton soon became redundant. The double-track line was first singled then through coaches to and from London were withdrawn, until in 1877 there was just one Parliamentary passenger train in each direction. Even this was withdrawn as a wartime measure in 1917 with complete closure coming in 1935.

The former MCR's line from Leicester (Wigston South) to Rugby was also relegated to branch line status in 1857 when the MR opened its new main line southwards from a junction at Wigston to Hitchin on the East Coast Main Line via Bedford. South of Hitchin MR trains were allowed to use the Great Northern Railway route to London King's Cross from February 1858. This arrangement did not last for long, ending in 1862 due to the overcrowded nature of the London terminus. Determined to reach London on its own metals the MR opened its new main line south from Bedford to St Pancras station in 1868. From this date the Bedford to Hitchin line was relegated to branch line status with much it being singled in the early twentieth century. Both the Wigston to Rugby and Bedford to Hitchin lines closed on 1 January 1962.

LEFT: Derby railway station on the North Midland Railway in 1840. The station was also jointly used by the Midland Counties Railway and the Birmingham & Derby Junction Railway, all constituent companies of the Midland Railway on its formation in 1844.

BELOW: Opening day of Nottingham railway station on the Midland Counties Railway in 1839. The station was built on Carrington Street in the city and acted as the terminus of the line.

The Railway Clearing House

Railways were spreading across the map of Britain by the early 1840s and at first sight it appeared that they were all linked, offering the opportunity of through passenger and goods traffic for the first time. But this was far from the truth as there was virtually no commonality between them. Despite most companies opting for the standard gauge of 4 ft 8½ in., with the exception of the Great Western Railway and its later constituent companies that chose Brunel's 7-ft 0¼-in. broad gauge – that is where the commonality usually ended.

Through carriages from one company to another were not allowed in many cases and some companies refused to carry third-class passengers from other lines. The through booking of a journey that involved using several railways was not possible either, there being no standardization of tickets. There was no standardization of goods vehicles, no common classification of goods and there were constant disputes over the allocation of receipts from goods traffic as there was no standardized distances between stations. Even station clocks had no commonly synchronized time.

In terms of safety, signalling was rather a hit-and-miss affair with some companies using red discs to indicate a 'clear road ahead' while others used the complete opposite! Whether privately owned or railway owned, goods wagons had different braking systems and even non-aligning buffers.

This all caused enormous confusion and was certainly a stumbling block to the continuing successful development of the railways. In 1842, painfully aware that some form of organization and standardization needed to be introduced, the Chairman of the London & Birmingham Railway invited representatives of eight other railway companies, many of them owned by the 'Railway King', George Hudson, to a meeting at Euston. Thus was formed the Railway Clearing House (RCH), although as membership was then voluntary some companies, notably the Great Western Railway, waited another twenty years before they joined.

The agreed RCH objectives were to organize the through booking of passengers, carriages and horses between different companies and to divide receipts based on mileage. The RCH also organized a classification of goods for goods traffic and tried to encourage a rate for these on a per-mile basis. Of course they were unable to set these rates as each company charged differently but it was a start. Member companies paid the RCH a fixed rate based on their number of stations and additionally a small percentage of their receipts.

The RCH had standardized railway timekeeping using Greenwich Mean Time in 1847 and before long had published distance tables, encouraged the use of Edmondson ticket-printing and date-stamping machines and had produced a general classification of goods. Inter-company debts were also settled at the RCH although these were not enforceable until 1850.

By 1850 a total of thirty-eight railway companies were making use of the RCH, being given legal status in the Railway Clearing Act in that year. Before that it had experienced considerable difficulty in carrying out its objectives because it did not have the power to prosecute or defend actions or suits, or take other legal proceedings. The railways that were parties to the clearing system were subject to the provisions of the act from 25 June that year, making inter-company debt collection easier for the RCH to enforce.

The RCH also published many railway maps, the first of which was produced by Zachary Macauley in 1851. John Airey continued his work, publishing his first Junction Diagrams in 1867. By the end of the nineteenth century his hand-coloured maps of the entire rail system, with distances shown between junctions and stations in miles and chains, were an important tool used by every railway company for calculating rates.

The Railway Clearing House had a staff of over 3,000 by the outbreak of the First World War, many of them employed in the painstaking task of number-taking to verify the numbers and contents of goods wagons travelling over the system – in fact by the time of the 'Big Four' Grouping in 1923 there were around 4,000 private owners using 650,000 goods wagons, and a sheet of paper had needed to be produced recording every journey made by each wagon – an enormous task! Additionally, the RCH checked passenger traffic returns of the different railway companies. Tickets surrendered to the collectors at the end of journeys were dispatched each month to the clearing house, a mammoth task on a railway system that had grown to 23,000 route miles by 1933.

The Railway Clearing House remained an important statistic-gathering organization until 1963 when, with much-reduced staff, it finally closed.

Thomas Edmondson and railway tickets

Born in Lancaster in 1792 Thomas Edmondson was a Quaker. After serving an apprenticeship as a cabinet-maker he went on to work for a local furniture company. He was appointed as stationmaster at Milton on the then uncompleted Newcastle & Carlisle Railway in 1836. One of his tasks at the station was to issue passengers with tickets, which at that time were laboriously handwritten on pieces of paper. Seeking to speed up this time-consuming process Edmondson devised a system of pre-printed cardboard tickets for each destination, which he numbered by hand and which were dated by a hand-operated machine. He further developed this by inventing a foot-operated date-stamping press until finally he patented a machine which printed tickets in batches, with each ticket being individually numbered – in this way railways could keep a record of tickets sold when they were surrendered at the end of each journey, thus vastly reducing fraud by unscrupulous railway staff who would often pocket fares for themselves.

The Newcastle & Carlisle Railway showed no interest in his invention, but in 1839 Edmondson was snapped up by the newly opened Manchester & Leeds Railway, where he was appointed chief booking clerk at Manchester Oldham Road station. In 1842 his ticket machine was adopted as a standard feature for its members' railways by the newly formed Railway Clearing House. Until his death in 1851 Edmondson went on to make a fortune from his patented ticket machine, charging railway companies an annual royalty of ten shillings (fifty pence) per mile of their routes. Also widely used on many European railways the system remained in use on parts of British Rail until 1990. Edmondson tickets can still be found in use on many heritage railways around the world, a fitting memorial to a humble cabinet-maker from Lancaster.

ABOVE: Introduced by the Manchester & Leeds Railway in 1839, Edmondson railway tickets and the machines that printed them made their inventor, Thomas Edmondson, a fortune and were widely used on European railways until the late twentieth century.

George Bradshaw and railway timetables

George Bradshaw was born in Pendleton, Salford, Lancashire, on 29 July 1801 and began his career working for J. Beale, a Manchester engraver. Bradshaw had become an accomplished cartographer after a short spell in Belfast and his first work, a map of his native county, was published in 1827. He published a series of maps of the canals of Yorkshire and Lancashire in 1830, known as *Bradshaw's Maps of Inland Navigation.*

In 1839 George Bradshaw published the world's first railway timetable in partnership with his London agent, William Adams. Sold with a cloth cover, *Bradshaw's Railway Time Tables and Assistant to Railway Travelling* (the title was shortened to *Bradshaw's Railway Companion* in 1840) was initially a fairly small publication of eight pages, costing 6*d.* (2½p), but as new railways quickly opened it grew in size until it had reached a staggering 1,000 pages by the end of the century. It was published monthly from 1841 under the new title *Bradshaw's Monthly Railway Guide* and included information on places to visit, hotels and shipping services, soon putting some order into the chaotic railway system.

From information supplied by the plethora of railway companies, Bradshaw compiled his updated timetable guide every month along with a map of Britain's rail system. Each line on the map was cross-referenced with the number of the page on which its timetable appeared. It was not an easy guide to use – there were so many companies that it was difficult to standardize the information and the multiplicity of footnotes and miniscule typography accompanying each timetable added more confusion. However, despite being the butt of Victorian cartoonists and music hall jokes, it became a unique national institution and was often referred to in famous works of fiction such as *Sherlock Holmes* and *Around the World in Eighty Days.* It was an invaluable tool for the travelling public for over a hundred years.

His name already synonymous with railway timetables, Bradshaw quickly expanded his railway publishing empire by introducing the *Continental Railway Guide* in 1847 (discontinued in 1939) and, later, the *Railway Manual* and the *Railway Shareholders' Guide,* both of which stayed in print until the early 1920s.

Bradshaw did not face any real competition until the 1930s, when the 'Big Four' railway companies started issuing their own timetables – apart from the Great Western Railway's, even these were produced by the publisher of *Bradshaw* until 1939. Despite this competition, and the publication of regional timetables by the nationalized British Railways after 1948, *Bradshaw* was the only all-line timetable. It continued to be available until 1961 when the final edition, No. 1521, was published.

It was 1974 before British Rail first published its own all-line timetable and this was last published in book form (by Network Rail) in 2007. Currently Network Rail publishes an Electronic National Rail Timetable that is updated twice a year. There are also numerous printed leaflets on a company-by-company or route-by-route basis that are available free of charge at stations – almost taking us back to the pre-Bradshaw situation.

Bradshaw & Blacklock, George Bradshaw's printing house, became internationally renowned for its maps, guides, books and the *Manchester Journal.* In his private life in his native Manchester, Bradshaw was a Quaker, a fervent peace activist and a philanthropist. While on a visit to Norway in 1853 he contracted cholera and died. He is buried in the grounds of Oslo cathedral.

In more recent times a facsimile edition of *Bradshaw's Handbook* of 1863 has become a bestseller in the UK following Michael Portillo's long running BBC TV series *Great British Railway Journeys.*

LEFT: Printer and engraver George Bradshaw published the world's first railway timetables.

RIGHT: With counties picked out in pastel colours this map of Britain's railways and sea routes was published by George Bradshaw in 1843.

BRADSHAW'S
Map of the Railways in Great Britain
shewing also the line of
NAVIGATION
from the principal SEA PORTS to both home and
Foreign Stations
1843
Scale of Miles
GERMAN
OCEAN
IRISH SEA
ST. GEORGES CHANNEL
CARDIGAN BAY
BRISTOL CHANNEL
THE ENGLISH CHANNEL
FRANCE
YORKSHIRE
LINCOLN
DURHAM
WESTMORELAND
ISLE OF MAN
ISLE OF MULL
Belfast
Longitude West
Longitude East

Government involvement and regulation

Private Acts of Parliament

As we have already discovered, the early wooden wagonways were built on an *ad hoc* basis on private land but in 1758 the Middleton Railway of Leeds became the first railway in Britain to be granted powers of permanence by an Act of Parliament. These were Private Acts, which passed powers or benefits to individuals or bodies rather than the general public. Parliament's role was to arbitrate between the promoters of the Acts and those affected by their projects – such as landowners – as well as to take account of the public interest. Such Acts might be required if a communal right of passage was being taken over or when land needed to be purchased and a joint-stock company created to raise capital. Unique to the United Kingdom at that time, these Private Bills meant that local transport needs could be brought before Parliament and a legislative tool provided, that in the case of railways allowed for the compulsory purchase of land. Unfortunately they could be very expensive if contested, and many were – the Great Northern Railway Bill of 1845–6 cost the promoters over £400,000.

Proceedings on railway Bills were not a foregone conclusion and many failed at their first or even second attempts. The Liverpool & Manchester Railway Bill of 1824 failed because George Stephenson was unable to provide adequate details of the costs and even the Great Western Railway Bill of 1834 failed despite fifty-seven days spent in committee. Parliament's decisions were not necessarily consistent and national interest was often neglected especially during the period in the 1840s known as 'Railway Mania' – proposals for overambitious and sometimes downright fraudulent schemes for new railways across Britain reached their peak in 1846 when 272 Acts of Parliament were passed.

1840 Railway Regulation Act

In 1840 the first step was taken by Parliament to regulate this fast-growing transport phenomenon. Given Royal Assent on 10 August 1840, the Railway Regulation Act included many measures aimed at bringing some kind of order out of the chaos. Resultantly Her Majesty's Railway Inspectorate was established to oversee safety on the railways. Board of Trade inspecting officers were responsible for investigating railway accidents and reporting findings and recommendations to Parliament. The first major accident investigated was that of a London to Bristol train on Christmas Eve 1841, which ran into a landslip in Sonning Cutting near Reading killing nine third-class passengers.

These feared men of the Railway Inspectorate also had the last say on whether a newly built line was ready and safe for opening – if not then railway companies faced delays and extra costs before starting public services.

1844 Railway Regulation Act

Until the passing of this Act the early railways provided three classes of carriages for passengers – the upper class sat in covered carriages with side-opening doors and windows, the middle class sat in carriages provided with roofs but open at the sides while the working class had to make do with a journey in an exposed wagon attached to a goods train, open to the elements and locomotive emissions. Fatal accidents in these open wagons were not uncommon, prompting Parliament to act. During 1844 a Parliamentary select committee had produced several reports on the railways for the Board of Trade – many of its proposals were distilled into this one Act.

The Railway Regulation Act of 1844 compelled all railway companies to provide cheap, basic travel for less-well-off working-class passengers, including provision of at least one train per day each way, average speeds of not less than 12 mph, stopping at all stations and in weather-protected carriages with seats. For this service railway companies could charge a maximum rate of one penny per mile – becoming known as 'parliamentary trains' they were normally laid on during unsociable hours.

While some railway companies provided the absolute minimum service required of them others, such as the Midland Railway, went much further in improving the lot of third-class passengers by providing separate compartments with windows and an oil lamp. In 1875 second-class coaches were abolished, being rebranded as third class. It wasn't long before other railway companies followed suit and from then until the mid-twentieth century passenger trains consisted of just first-class and third-class carriages.

1846 Regulating the Gauge of Railways Act

The Great Western Railway Act finally passed by Parliament in 1835 had one serious flaw – it did not specify the gauge of the railway. Britain's other new railways were sensibly adopting George Stephenson's 4-ft 8½-in. gauge as the industry standard but the Great Western's engineer, Isambard Kingdom Brunel, dismissed this as a gauge used by colliery lines. He opted for a wider gauge of 7 ft 0¼ in. for his new high-speed railway between London and Bristol which opened in 1841. This lack of standardization was also felt in Ireland, where by 1845 railways were using a confusing mixture of 4-ft 8½-in., 5-ft 3-in. and 6-ft 2-in. gauges.

A Royal Commission was set up by Parliament to look into the vexed problem of standardizing rail gauges and their findings formed the basis of the Regulating the Gauge of Railways Act which came into force on 18 August 1846. The Act mandated all new passenger railways in Britain to be built to the standard gauge of 4 ft 8½ in. while Ireland's standard gauge was set at 5 ft 3 in. The only exceptions were the railways of southwest England and South Wales where Brunel's broad gauge had already become established. In effect the Act spelt the end for the broad-gauge system although it took until 1892 before the GWR had completely converted its network to the standard gauge.

LEFT: Another accident to be investigated by the Railway Inspectorate – ten passengers were killed and forty-nine injured when a South Eastern Railway boat train derailed on the bridge over the River Beult near Staplehurst in 1865. Amongst the survivors was the author Charles Dickens.

RIGHT: 'A Day at the Races' – a contemporary engraving from the Illustrated London News in 1846 shows the social distinctions through the division of rail travel into first, second and third class.

Isambard Kingdom Brunel

A giant of a man ahead of his time, a visionary and innovative engineer, Isambard Kingdom Brunel was born in Portsmouth in 1806. After moving with his parents to London at the age of two, Isambard was educated at home by his father, French-born engineer Marc Isambard Brunel, and at a young age had already mastered the techniques of drawing buildings, geometry and basic engineering principles. Later he received a first-class education in France, studying at the Lycée Henri-IV in Paris before training as a clockmaker with the Breguet company.

Returning to Britain in 1822, Isambard worked with his father on the construction of the Thames Tunnel between Rotherhithe and Wapping, in which he narrowly escaped drowning during a flood. The tunnel is still in use today as part of the London Overground system.

At the tender age of 25 Brunel won a competition to build the Clifton Suspension Bridge in Bristol but work on it was suspended by the Bristol Riots, which drove away investors. This graceful bridge was not completed until 1864, five years after his death.

Brunel was appointed Chief Engineer of the fledgling Great Western Railway (GWR) in 1833. Parliamentary approval for a railway between London and Bristol came in 1835 but his choice of a broad gauge of 7 ft 0¼ in. set the GWR apart from the rest of Britain's expanding rail network. Nevertheless this superbly engineered and near-level line opened throughout in 1841 and soon became known for its speed and comfort. Despite his success with the GWR Brunel appears to have made the wrong decision when choosing an atmospheric railway system for a line in South Devon. It had a short working life from 1847 to 1848 before being replaced by conventional locomotive-hauled trains. However, Brunel's unique broad gauge lasted much longer, only being eradicated in 1892 when the GWR completed its conversion to standard gauge.

Without a doubt Brunel was a visionary and saw the GWR as the first stepping stone for a transatlantic route between London and New York. To further this dream he had been involved in the setting up of the Great Western Steamship Company in 1835. Designed to provide a fast and comfortable service between Bristol and New York Brunel's wooden paddle-wheel steamship, the 1,340-ton *SS Great Western*, was launched at Bristol in 1837. She was the first steamship designed to carry passengers across the Atlantic and made her successful maiden voyage in 1838, remaining in service on this route until 1847.

SS Great Western was replaced by Brunel's *SS Great Britain*, the first iron-hulled screw steamship to cross the Atlantic. She was launched in 1843, making her debut transatlantic crossing in 1845. However, the ship foundered a year later and the cost of refloating her bankrupted the Great Western Steamship Company. She was later converted to an emigrant ship carrying 750 passengers, making thirty-two return voyages to Australia until 1876. After spending many years as a hulk at the Falkland Islands she was returned to Bristol in 1970 where, now restored, she is a popular tourist attraction.

Brunel's final steamship was a monster. By far the biggest ship in the world when launched in 1858, the 19,000-ton *SS Great Eastern* was designed to carry 4,000 emigrants to North America. Sadly her maiden voyage in

1859 was a disaster – while sailing to Weymouth on 9 September a boiler exploded killing five crewmen and throwing one of the five funnels clear of the deck. Aged 53, Brunel died from a stroke just six days later. His leviathan ship was repaired and spent her latter years laying over 30,000 miles of undersea cables around the globe before being broken up in 1890.

While his failures have long ago disappeared, Brunel's many railway achievements live on into the twenty-first century. In addition to his magnificent legacy of a super-smooth iron highway between Paddington and Bristol (soon to be electrified) via Box Tunnel, other highlights which are still in regular use today are Maidenhead Railway Bridge over the Thames in Berkshire, the Royal Albert Bridge over the Tamar near Plymouth (completed just before his death), the highly scenic Cornish railway route from Plymouth to Penzance and the equally scenic line between Exeter and Plymouth, his South Wales' trunk route and the Cotswold Line between Oxford and Worcester. Of note on the latter is Campden Tunnel where Brunel had a spot of bother with a contractor during its construction – apparently IKB was not an easy man to work for! And while in Bristol do walk across his magnificent Clifton Suspension Bridge over Avon Gorge and visit the restored *SS Great Britain.*

LEFT: Isambard Kingdom Brunel poses for the camera in front of the enormous anchor chains of his leviathan of the seas, the *SS Great Eastern.*

BELOW: Taken in 1858 this photograph by Roger Fenton – famous for his photographs of the Royal Family and the Crimean War – shows Brunel's last masterpiece, the Royal Albert Bridge over the River Tamar near Plymouth, under construction.

Atmospheric railways

Isambard Kingdom Brunel turned his attention to his next railway project, the 52-mile broad-gauge South Devon Railway (SDR) from Exeter to Plymouth following the opening of the broad-gauge Bristol & Exeter Railway (B&ER) in 1844. Planned to take a fairly level coastal route to Teignmouth, this new railway would then encounter severe gradients as it wound its way across the southern slopes of Dartmoor before reaching Plymouth. After witnessing a demonstration of the London & Croydon Railway's atmospheric system, Brunel argued that the system would be capable of moving heavier loads over the difficult terrain of the South Devon Banks than could at that time be achieved with adhesion steam locomotives. The South Devon Railway received Parliamentary approval just two months after the Bristol & Exeter had opened, receiving financial backing from the Great Western, the Bristol & Exeter and the Bristol & Gloucester railways.

Dispensing with locomotives altogether, the system centred around a slotted iron pipe laid between the tracks – inside the pipe was a piston, connected to the underside of the leading coach by an iron rod, which was sucked along by a permanent vacuum created by steam-driven pumping houses set at 3-mile intervals along the route. In theory the slot in the top of the pipe was sealed with air-tight leather flaps, which first opened and then closed as the train progressed.

British engineers the Samuda brothers had patented the atmospheric railway system in 1844. Two other atmospheric railways were already operating in Europe in addition to the London & Croydon's line and Brunel's proposed South Devon Railway. The first to open was the 1¾-mile Dalkey Atmospheric Railway, an extension of the Dublin & Kingstown Railway in Ireland, which successfully operated between 1844 and 1854 before being converted to conventional haulage. In France between 1847 and 1860 a 1-mile extension of the Paris-St Germain Railway was operated using the atmospheric system.

In practice, however, the atmospheric system proved to be a resounding failure as the leather flaps were completely unsuccessful in ensuring a permanent air-tight fit – not only did rats find the leather very tasty but during warmer weather the flaps dried and cracked. As Brunel's route closely followed the coastline the salt spray from the sea also created havoc with the leather flaps. The South Devon Railway had already commenced operations before these major problems were fully encountered, however – Teignmouth had been reached in September 1847 and Totnes by July 1848. With mounting maintenance costs taking their toll, the South Devon Railway's concerned shareholders stepped in and the system ceased to operate on 9 September 1848, being replaced by hired-in conventional steam haulage.

The rest of the route between Totnes and Plymouth had already opened as a conventional railway in May of that year, using locomotives hired from the Great Western Railway (GWR). The SDR was amalgamated with the GWR in 1878 and, having already swallowed up the B&ER two years earlier, this much-enlarged company now controlled the entire route from Paddington to Plymouth and thence to Cornwall. In 1892 Brunel's broad gauge disappeared entirely when the route west of Exeter to Penzance was converted to standard gauge in the space of just one weekend.

Three of the Italianate-style pumping engine houses survive today: at Starcoss adjacent to the railway; at Totnes adjacent to the station – this was never brought into use; and at Torquay – never used as the branch from Newton Abbot to Torquay was built after the abandonment of the atmospheric system.

LEFT: Three views of Brunel's atmospheric system on the South Devon Railway. The top illustration shows construction of the line with the slotted iron atmospheric pipe being laid in the middle of the track. In the centre the railway is seen on its coast-hugging route near Dawlish while the lower view shows one of the pumping houses that were built at 3-mile intervals along the line.

ABOVE: A section of the South Devon Railway's atmospheric system can be seen today at the Didcot Railway Centre.

Battle of the gauges

When the Great Western Railway received Parliamentary authorization on 31 August 1835, there was surprisingly no stipulation about the gauge of the line (the width between the rails) – nearly all British railways were by then adopting the standard gauge of 4 ft 8½ in. introduced by George Stephenson for the Stockton & Darlington Railway and Liverpool & Manchester Railway. Seizing upon this glaring omission, Brunel, who had been appointed chief engineer of the GWR in 1833, successfully advocated a broad gauge of 7 ft 0¼ in. The gauge offered more stability for trains, allowing journeys between the two cities to be completed at unheard-of speeds and in sumptuous comfort.

In its early years the broad-gauge Great Western Railway was a resounding success and heralded a new age of high-speed inter-city rail travel. Indeed, by 1867 the GWR's tentacles, with the help of its neighbouring broad-gauge supporters, had reached Penzance with unprecedented journey times of nine hours from London – today it takes just over five hours. Unfortunately the rest of the country's railways were being built to the narrower, standard gauge of 4 ft 8½ in. and where these railways came into contact with the GWR, the scenes of confusion were recorded for posterity by Victorian cartoonists. Probably the most famous of these cartoons depicts the farcical scenes at Gloucester where the standard-gauge Midland Railway met the broad-gauge GWR. Chaos reigned as passengers were forced to change trains while goods had to be manhandled from broad-gauge wagons to narrow-gauge wagons and vice versa. A public meeting was held in 1844 where Birmingham businessmen protested at the delays as 'a commercial evil of the first magnitude'.

A Royal Commission was set up to look into the vexed question of railway gauges but before the Railway Gauge Act of 1846 could be implemented the GWR had received Parliamentary approval for its new trunk route through South Wales to be built to the broad gauge. The Act, which came into force on 18 August 1846, made it illegal to build passenger railways in Britain on any gauge other than 4 ft 8½ in. The only exceptions to this were the railways of southwest England and South Wales where Brunel's broad gauge had already become established. Even so the GWR was forced to use mixed-gauge track on its Oxford to Birmingham route – both broad-gauge and standard-gauge trains could now use the route.

Following Brunel's death on 15 September 1859, the individualistic GWR soon came to the inevitable conclusion that the broad-gauge dream was incompatible with the rest of the nation's railways. Mixed-gauge track soon began to spread on many of the GWR's routes and in 1861 standard-gauge trains started operating between Paddington and Birmingham. Starting in 1864, broad-gauge track was slowly converted to standard gauge and in many places mixed-gauge track was common for some years. By 1892 all that remained of the broad gauge was the main line from Paddington to Penzance and this was converted by an army of workmen over one weekend in May of that year. George Stephenson's standard gauge had won, not only in Britain but it also went on to become the standard for the majority of railways in the world.

LEFT: This working replica of Daniel Gooch's 4-2-2 locomotive *Iron Duke* runs on 7-ft 0¼-in.-gauge track. The original was built by the GWR in 1847 and was capable of speeds of up to 80 mph.

TOP RIGHT: This cartoon shows the chaos that ensued at Gloucester station in 1846 where the broad-gauge GWR met the standard-gauge Midland Railway, forcing passengers on their way to Birmingham to change trains.

BOTTOM RIGHT: The GWR's broad-gauge system was finally converted to standard gauge in 1892. Here nearly 200 broad-gauge locomotives are seen stored at Swindon Works, awaiting either scrapping or conversion to standard gauge.

TO THE
BIRMINGHAM TRAIN
GLASS

Chester & Holyhead Railway

Since 1546 the Irish Mail had been carried between London and Dublin over rutted roads by horse and stagecoach to Holyhead and then by sailing ship over the stormy Irish Sea. This important route was vastly improved during the early nineteenth century when Thomas Telford upgraded what is now known as the A5 road across Snowdonia and built the Menai suspension bridge across the Menai Strait to Anglesey. Despite all of these improvements and the introduction of steamships for the sea crossing, it was still a slow and lengthy journey by stagecoach – London to Holyhead alone took thirty hours – and by the 1830s it was no match for the new railways which were already spreading across Britain. The completion of the railway route between London and the north in 1838 via the London & Birmingham, Grand Junction and Liverpool & Manchester railways temporarily brought an end to Holyhead's importance. From early 1839 the Irish Mail service was rerouted to run by train to and from Liverpool, from where the mail was carried by fast steam packets, reducing the London to Dublin journey time to 22½ hours

Chester had been reached in 1840 by the opening of the Chester & Crewe Railway and it wasn't long before proposals were put forward for a new railway along the North Wales coast and across Anglesey to Holyhead. Financially backed by the London & Birmingham Railway, the Chester & Holyhead Railway (C&HR) was incorporated in 1844 but the decision on how to cross the Menai Straits was delayed until the following year. Engineered by Robert Stephenson, the railway was opened in two separate sections in 1848, namely Chester to Bangor and across Anglesey to Holyhead. On the former section of line Stephenson had used a novel, but costly, type of bridge – built of rectangular wrought-iron tubes through which trains could pass – to carry the line over the River Conway.

The intervening gap across the Menai Strait was also crossed by a tubular bridge – named the Britannia Tubular Bridge – which opened in 1850. With the completion of the railway the Irish Mail was once again routed via Holyhead, a service which continued without interruption until 1985. To cater for the seagoing traffic between Holyhead and Dun Laoghaire (for Dublin) the C&HR put into service four new ships, increasing to nine by 1856.

While Stephenson's tubular bridges were a great success the same cannot be said of his bridge over the River Dee in Chester. Built with cast-iron beams

strengthened by wrought-iron tie bars, it collapsed in 1847 just as a train destined for Ruabon was crossing – the engine and its driver escaped but the tender and passenger coaches fell into the Dee killing the fireman, the guard and three passengers. The use of cast iron in the building of bridges came to a dramatic conclusion in 1876 with the collapse of the Tay Bridge in Scotland.

The building of the Chester to Holyhead Railway had left the C&HR in dire financial straits and the line was worked from the outset by the newly formed London & North Western Railway (LNWR). Despite the growth in traffic along the line, including through expresses to and from Euston and the carrying of considerable amounts of coal, slate and Irish cattle, the company remained unprofitable and in 1858, somewhat to the shock of the LNWR, approached the Great Western Railway for help. Not surprisingly, the LNWR's headquarters at Euston could not countenance this encroachment into their territory and made a counter offer which was accepted by the C&HR. The company was absorbed by the LNWR at the beginning of 1859.

The opening of the railway not only saw the renaissance of Holyhead (where the Great Breakwater was completed in 1870) as the premier port for Ireland but also brought about a massive development of resort towns along its route. Prestatyn, Rhyl, Colwyn Bay and Llandudno all owe their popularity as destinations for Victorian and twentieth-century holidaymakers to the railway. Llandudno itself was connected to the mainline in 1858 with the opening of a 3-mile branch from Llandudno Junction station.

The Chester to Holyhead line along the North Wales coast and across Anglesey remains in operation today with diesel trains from London Euston, Manchester, Birmingham and Cardiff meeting ferries to and from Dublin and Dun Laoghaire in Ireland. Sadly Stephenson's Britannia Tubular Bridge was destroyed by fire in 1970 but was rebuilt as a 2-tier road/rail bridge that reopened to rail traffic in 1972.

LEFT: Overlooked by thirteenth-century Conwy Castle, Robert Stephenson's 410-ft-long tubular bridge was built in 1849 and carried the Chester & Holyhead Railway across the River Conwy.

BELOW: Seen here under construction in 1849, Stephenson's largest tubular bridge carried the Chester & Holyhead Railway across the Menai Straits to Anglesey. The bridge was destroyed in a fire in 1970 but subsequently reopened as a 2-tier road/rail bridge.

Eastern England

What was to become the Great Eastern Railway's mainline from Liverpool Street to Norwich started life as the Eastern Counties Railway (ECR). Incorporated in 1836, the 126-mile line from London to Great Yarmouth via Norwich was the longest railway in Britain to be authorized by Parliament at that time. However, the company's construction costs were grossly underestimated and by the time the line had reached Colchester these had amounted to 50 per cent more than the original budget estimated for the whole route.

At a time when Parliament did not mandate the rail gauge, the railway was built to a gauge of 5 ft although it was converted to standard gauge only a year after opening. The first section from a temporary terminus in Mile End to Romford opened in 1839, from Mile End back to the new terminus at Bishopsgate and from Romford to Brentwood in 1840 and on to Colchester in 1843.

Meanwhile the 20½-mile Yarmouth & Norwich Railway (Y&NR) had been authorized in 1842. Surveyed by George and Robert Stephenson, this level route follows the meandering Yare Valley and crosses reclaimed marshland to the coast, the only major engineering feature being a ½-mile diversion of the river to the east of Norwich, which was considered cheaper than building two bridges.

The first railway to be built in Norfolk, the Y&NR opened in 1844 and was amalgamated with the newly opened Norwich & Brandon Railway the following year to form the Norfolk Railway. The arrival of the railway in Great Yarmouth soon breathed life into the ailing port and by the late nineteenth century the enlarged and modernized harbour was home to 1,000 trawlers, their vast catches of herring and other fish being rapidly transported by train to distant markets. By this time, with rail connections from London, the Midlands and the North now in place, the town and its long sandy beaches had also become one of the most popular holiday destinations in Britain. The line became so congested

that a new relief railway was opened in 1883 between Brundall and Breydon Junction, one mile west of Great Yarmouth, via the town of Acle.

While the Eastern Counties Railway initially failed to reach Norwich and Great Yarmouth, it did eventually achieve its goals through various takeovers, leasing and working arrangements. Firstly, in 1844, it leased the Northern & Eastern Railway, which had already opened its 5-ft-gauge line from Islington to Bishop's Stortford in 1842. This, too, was converted to standard gauge and extended to Cambridge and Brandon in 1845. The ECR then took over the Norfolk Railway in 1848 (see above), absorbed the bankrupt East Anglian Railways in 1852 and took over working the Eastern Union Railway in 1854. The latter had already opened its line throughout from Colchester to Norwich (Victoria) and from Ipswich to Bury St Edmunds in 1849. Despite its faltering start only thirteen years previously, the Eastern Counties Railway now controlled much of the railway network in East Anglia but soon, deservedly, acquired a reputation for its poor service

In 1862 the ECR became the main constituent company of the newly formed Great Eastern Railway and services soon started to improve. The former ECR works at Stratford was expanded as major locomotive works and a new London terminus at Liverpool Street, built on the site of Bethlem Hospital (or 'Bedlam' as it was then known), was completed in 1874.

Also opened in 1849, the ECR's Victoria station in Norwich remained the terminus of passenger services from London until 1886 when most of them were diverted to the Great Eastern Railway's grand new Thorpe station. Victoria station closed to passengers in 1916, Thorpe station is still open today while the Midland & Great Northern Joint Railway served Norwich's third terminus, City station, from 1882 to 1959.

LEFT: Construction of the Eastern Counties Railway near Ilford in Essex in 1838. The railway opened between Shoreditch and Colchester in 1843, but it was left to the Eastern Union Railway to extend the line to Norwich.

BELOW: The Norfolk Railway was one of the first in Britain to use a railway telegraph signalling system. Made by W.F. Cooke, this block signalling instrument was introduced in 1845 and gave a complete picture of the traffic between Norwich and Brandon.

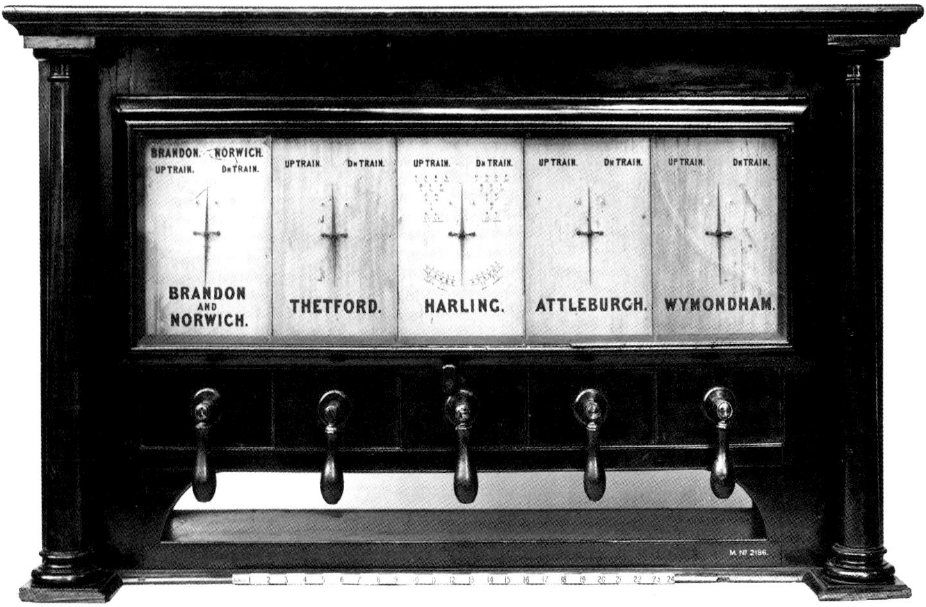

London & North Western Railway

One of Britain's, if not the world's, most successful railway companies of the late nineteenth century started life on 16 July 1846 with the amalgamation of the London & Birmingham Railway, the Grand Junction Railway and the Manchester & Birmingham Railway. The amalgamation was prompted by the Great Western Railway's proposal to extend its broad-gauge network northwards to Birmingham. Known as the London & North Western Railway (LNWR) the new company initially controlled 350 route miles of lines connecting London (Euston) with Rugby, Birmingham, Wolverhampton, Crewe, Chester, Liverpool and Manchester.

However, by the gradual acquisition of over sixty other railway companies the LNWR eventually controlled over 2,500 route miles of railways linking London not only with the Midlands but also South Wales, North Wales and northwest England. Among these routes were the important links with Ireland and Scotland; the former via Holyhead and the latter via the West Coast Main Line to Carlisle and thence over Caledonian Railway metals to Glasgow and Edinburgh.

Advertising itself as the 'Premier Line', the railway introduced many innovations such as water troughs that enabled steam locomotives to collect water while travelling at speed (1860) and electrification on its London commuter routes in the early twentieth century. The company also operated a fleet of ships on its Irish Sea routes from Holyhead. In Birmingham, Britain's second biggest city, the LNWR opened a new station in 1854 to replace the original London & Birmingham terminus at Curzon Street – at that time New Street station then had the largest single-span iron and glass arched roof in the world.

At Crewe the LNWR greatly expanded the railway works established there in 1840 by the Grand Junction Railway. In 1853 the works started manufacturing rails and in 1864 opened its own steelworks. A railway town soon became established with all the needs of its workers and families being looked after by a paternalistic LNWR. At its peak Crewe Works employed over 20,000 people while the company as a whole employed well over 100,000. In 1923 the LNWR, then with a route mileage of 2,667½ miles, was the largest of seven major railway companies that constituted the new London Midland & Scottish Railway. Today the former mainlines of the LNWR are all still in business, many of them now electrified and providing important arteries for long-distance passenger and freight services.

BELOW: This romantic view of a London & North Western Railway train crossing a viaduct at Newton in Cheshire was lithographed by American-born artist Arthur Fitzwilliam Tait in 1848.

RIGHT: Water troughs were first introduced in Britain by the LNWR in 1860. Here unique LNWR 2-4-0 locomotive *Jeanie Deans* is seen at speed picking up water at Bushey Troughs in 1899.

Completion of the West Coast Main Line

The southern part of the West Coast Main Line as we know it today was beginning to take shape in 1838 upon completion of the London & Birmingham Railway (L&BR). From 17 September that year it was possible to travel by rail from London Euston to Manchester or Liverpool without changing trains at Birmingham's Curzon Street station – north of here Liverpool or Manchester were reached via the Grand Junction Railway (GJR) which had opened a year earlier.

With the opening of the Trent Valley Railway (TVR) between Rugby and Stafford in 1847, journey times between London and the North were considerably reduced. By this date the TVR had already become part of the London & North Western Railway that had been formed by the amalgamation of the L&BR, GJR and the Manchester & Birmingham Railway in 1846.

Four different railway companies were responsible for building the middle section of the West Coast Main Line between Crewe and Carlisle, all of which, apart from one, had been absorbed by the London & North Western Railway (LNWR) by 1879.

In 1838 the North Union Railway (NUR) opened the West Coast Main Line north of Newton Junction to Preston. The GJR and the Manchester & Leeds Railway later leased it jointly and their successors, the LNWR and Lancashire & Yorkshire Railway respectively, continued with this arrangement until the 'Big Four' Grouping of 1923 upon becoming part of the LMS. To the north of Preston the Lancaster & Preston Junction Railway (L&PJR) opened in 1840. This line was first worked by the NUR but the arrangement ended in acrimony in 1842 and the L&PJR, in a fit of spite, leased itself to the Lancaster Canal Company. This arrangement continued until 1849, when it was leased by its new northern neighbour, the Lancaster & Carlisle Railway (L&CR), who went on to take over the L&PJR ten years later.

The 69-mile Lancaster & Carlisle Railway, the final section of the major trunk route from London to Carlisle, was authorized in 1844. Engineered by Joseph Locke and

built by Thomas Brassey the railway took a route through the Lune Gorge before ascending the final three miles on a ruling gradient of 1 in 75 from Tebay to Shap Summit. Shap was approached from the north along a 10-mile climb on a gradient of 1 in 125. It opened in 1846 after taking only 2½ years to complete and employing 10,000 navvies. Worked from the outset by the newly formed LNWR, it remained independent until 1879 when it was taken over by the latter.

Joseph Locke surveyed in 1836 the final part of what became known as the West Coast Main Line. The Caledonian Railway (CR), which became one of Scotland's most powerful railway companies, was authorized by Act of Parliament in 1845 to build a main line along Locke's route between Carlisle and Glasgow together with a junction at Carstairs for a branch to Edinburgh.

The new railway opened from Carlisle to Beattock on 10 September 1847, and from Beattock to Edinburgh and Glasgow on 15 February 1848. A connection with the Scottish Central Railway at Castlecary was opened on 7 August 1848, with a new terminus opening at Glasgow's Buchanan Street in 1849. The Caledonian Railway lost no time in making full use of its rail connection with England and had soon usurped the East Coast Main Line (which still required change of trains at Newcastle and Berwick) for the carrying of mail and passengers, becoming the first railway to convey through carriages between London and Scotland. Buchanan Street station in Glasgow soon proved inadequate to cope with the increased traffic and in 1879 the company opened Central station north of the Clyde. In turn, between 1901 and 1905, Central station was later enlarged to its present grandeur.

With its close working relationship with the LNWR south of the border, the Caledonian Railway's (Caley) route between Carlisle and Glasgow saw the introduction of the famous express 'The Royal Scot' between London Euston and Glasgow as early as 1862. By the end of the nineteenth century this 10 a.m. departure from Euston was being hauled north of Carlisle behind McIntosh's legendary 'Dunalastair' 4-4-0s, built at the CR's own Glasgow St Rollox Works.

FAR LEFT: Built by the Trent Valley Railway in 1846 in a style sympathetic to the nearby Earl of Lichfield's stately home, Shugborough Tunnel is still used today by trains on the West Coast Main Line.

LEFT: Civil engineering contractor Thomas Brassey was responsible for building much of the world's railways in the nineteenth century, among them the Trent Valley Railway, the Lancaster & Carlisle Railway and the Caledonian Railway, all of which went on to form the West Coast Main Line.

RIGHT: The West Coast Main Line in all its glory is depicted in this poster produced for the LNWR and Caledonian Railway.

1851	1852	1854	1859	1860	1863	1863	1868	1871	1873	1876
	Great Northern Railway between King's Cross and Doncaster opens	North Eastern Railway formed	Brunel's Royal Albert Bridge opens	First water troughs introduced by LNWR	Metropolitan Railway opens in London	Steam locomotives introduced by narrow-gauge Ffestiniog Railway	Midland Railway opens St Pancras Station in London	George Hudson dies a broken man	First sleeping car introduced by LNWR	Settle-Carlisle Line opened by Midland Railway

THE GREAT VICTORIAN RAILWAY AGE

1851–1900

1878	1879	1881	1886	1887	1890	1892	1894	1895	1896	1899	1900
First Tay Bridge opens	Tay Bridge collapses killing 75 people	First all-Pullman train introduced by LB&SCR	Severn Tunnel opened by GWR	Second Tay Bridge opens	Forth Bridge opens	End of GWR broad gauge	West Highland Railway opens between Glasgow and Fort William	Culmination of 'Railway Races to the North'	Light Railways Act	Great Central Railway London Extension opens	

East Coast Main Line

By 1840 it was possible to travel by train from London to York, albeit via a time-consuming and indirect route via Hampton-in-Arden, Derby and Leeds. Much of the journey was over the railways of George Hudson, the wealthy but corrupt Mayor of York, railway financier and later politician who became known as 'The Railway King'.

London (King's Cross) to York

Addressing this undesirable situation, the newly formed Great Northern Railway was authorized in 1846 to build a direct line from London to York via Peterborough and Doncaster with a loop line from Peterborough to Retford via Lincoln. By that date Hudson controlled over 1,000 miles of railway stretching from Birmingham to Newcastle and was infuriated by the new scheme, which threatened his railway empire.

Despite Hudson's underhand tactics attempting to delay the new railway, William Cubitt was appointed chief engineer for the fledgling GNR. Work started on the London to Peterborough section with Thomas Brassey acting as main contractor. Due to the terrain north of London work was slow, with two viaducts at Welwyn and nine tunnels required. Meanwhile, the Peterborough to Retford loop line via Lincoln was forging ahead across flat lands, opening in 1848 as far as Lincoln and to Doncaster via Retford in 1849.

In 1852 the railway opened from Peterborough to King's Cross. In the same year the direct mainline north of Peterborough opened to Retford and the GNR mainline to Shaftholme Junction north of Doncaster was complete. Until 1871 through running to York was only achieved by running powers between Burton Salmon and Knottingly over the York & North Midland Railway and the Wakefield, Pontefract & Goole Railway. Matters greatly improved in 1871 when the North Eastern Railway opened a direct line (via Selby) from Shaftholme Junction, north of Doncaster, to Chaloner's Whin Junction, on the Leeds line south of York. The East Coast Main Line to York was now in place and the route remained unchanged until the opening of the Selby cut-off in 1983.

In 1853 GNR established its locomotive, carriage and wagon works at Doncaster – under successive chief mechanical engineers, Patrick Stirling, Henry Ivatt and Nigel Gresley, it produced some of the finest steam locomotives in the world.

GNR traffic from King's Cross proved slow to develop due to Hudson's and his successor's tactics. By 1860 this had all been resolved, the GNR by then enjoying a massive surge of traffic along its main line, with through trains between King's Cross and Edinburgh. The three companies involved, the GNR, the North Eastern Railway and the North British Railway, developed special vehicles

for this service known as East Coast Joint Stock. The 10 a.m. 'Special Scotch Express' departure from King's Cross to Edinburgh, introduced in 1862, was the forerunner of the LNER's 'Flying Scotsman'.

York to Newcastle

The Great North of England Railway (GNoER) built the line between York and Newcastle, initially obtaining authorization in 1836 to construct a railway from Newcastle to Croft, south of Darlington, where it was to meet a branch of the Stockton & Darlington Railway. A year later the GNoER was authorized to continue the railway southwards to York. Here it was proposed to connect with the planned York & North Midland Railway (Y&NMR), one of the many railways being financed at this time by the 'Railway King', George Hudson.

The southern end of the line from Darlington to York presented no major engineering problems and opened in 1841 – the northern end from Darlington to Newcastle was not so straightforward. Despite receiving authorization to build the line in 1837, the GNoER reneged and in 1843 the incomplete section north of Darlington was taken over by Hudson's newly incorporated Newcastle & Darlington Junction Railway (N&DJR), opening in 1844. The railway possessed no rolling stock or locomotives and had to initially lease these from the GNoER. In spite of this setback the N&DJR prospered, taking over several other northeastern railways, changing its name to the York & Newcastle Railway (Y&NR) in 1846.

A year later the Y&NR amalgamated with the Newcastle & Berwick Railway which had just opened its line northwards from Newcastle to Tweedmouth, becoming the York, Newcastle & Berwick Railway (YN&BR) – the East Coast Main Line as we know it today was gradually taking shape.

Newcastle to Edinburgh

While the final section of the Carlisle to Glasgow West Coast Main Line was forging ahead, the Tyne and Tweed rivers hampered construction of the rival East Coast Main Line between Newcastle and Edinburgh. Hudson's Newcastle & Berwick Railway (N&BR) received authorization in 1845 and the new railway struck northwards from existing lines at Heaton to Tweedmouth, south of Berwick. It opened throughout on 1 July 1847, immediately merging with Hudson's York & Newcastle Railway to form the York, Newcastle & Berwick Railway.

PREVIOUS SPREAD: The opening of the Forth Bridge in 1890 completed the East Coast Main Line route between London King's Cross and Aberdeen. Seen here in 1895, the Channel Squadron of the Royal Navy steams up the Firth of Forth led by Captain Bromley's flagship, *HMS Blake*.

LEFT: The London terminus of the East Coast Main Line was opened by the Great Northern Railway in 1852. In this scene Queen Victoria is arriving at King's Cross station in 1853 en route to York Races.

BELOW: Opened in 1841 inside the city walls, the original station at York was the terminus of the York & North Midland Railway and the Great North of England Railway. As trains were required to reverse out before continuing their journey it was replaced by the current through station in 1877.

The crossing of the two rivers took longer. Designed by Robert Stephenson, the High Level Bridge across the Tyne at Newcastle was opened by Queen Victoria in 1849, its 1,337-ft span also carrying a road on a lower level. Less than a year later the Queen opened John Dobson's fine Central station but, until the opening of the King Edward VII Bridge in 1906, all trains were forced to reverse before resuming north or south journeys. The missing link over the Tweed, Robert Stephenson's iconic Royal Border Bridge, was also opened by Queen Victoria in 1850. With the North British Railway having already opened the line north of Berwick to Edinburgh via Burnmouth and Dunbar in 1846, through trains could finally run between King's Cross and Edinburgh.

BELOW: Designed by Robert Stephenson and opened in 1849, the High Level Bridge was built to carry the York, Newcastle & Berwick Railway 120 ft above the River Tyne to the new Newcastle Central station.

RIGHT: Seen here in an LNER poster, the Royal Border Bridge carries the East Coast Main Line high across the River Tweed at Berwick. The 28-arch bridge was designed by Robert Stephenson and opened by Queen Victoria in 1850.

SCOTLAND
EAST COAST ROUTE

Brunel's broad gauge in South Wales

The broad-gauge South Wales Railway (SWR), backed by the Great Western Railway (GWR), was seen as the western half of a major trunk route linking London to Fishguard, where sea connections could be made to Ireland. Also included in this scheme were branches from Whitland to Pembroke and from Clarbeston Road to Neyland. The GWR had already reached Gloucester via Swindon in 1845 and the 172½-mile extension, financially backed by the GWR, gained Royal Assent in the same year. The railway was surveyed and engineered by Isambard Kingdom Brunel, following a very level route down the west shore of the Severn Estuary to Chepstow, Newport and Cardiff then westwards to Swansea. West of Swansea the only challenging gradient is the 1-in-52 climb up to Cockett Tunnel between Landore and Gowerton.

The railway opened in stages: Chepstow to Swansea in 1850; Swansea to Carmarthen and Gloucester to Chepstow in 1852; to Haverfordwest in 1854; and to Neyland (for Milford Haven) in 1856. And there the SWR stopped – the branch to Pembroke was not built, neither was the line to Fishguard Harbour or the steamer service to Rosslare in Southern Ireland, the latter only opening for business in 1906.

However, Brunel's superbly engineered but truncated route had one serious drawback – it was built to his unique broad gauge of 7 ft 0¼ in. while the majority of the other railways that served the South Wales' coalfields were built to the standard gauge of 4 ft 8½ in. Parliamentary approval for the line was granted just one year before the outlawing of new broad-gauge lines in the 1846 Regulating the Gauge of Railways Act. The SWR was soon paying dearly for its different gauge, losing out on lucrative coal traffic that was being sent by the standard-gauge lines direct to South Wales' ports for onward sea transshipment. Missing out on both Irish and coal traffic, the SWR was soon in trouble and was amalgamated with the GWR in 1863. Sense eventually prevailed at the GWR's Paddington headquarters and the whole route was converted to standard gauge in 1872, twenty years before the end of the broad gauge in southwest England.

Meanwhile in 1859 another company, the Pembroke & Tenby Railway, had been authorized to construct a line between Pembroke Dock and the expanding seaside resort of Tenby. Although this opened in 1864, the line had no physical connection with the rest of the railway network and remained isolated until 1866 when it was extended northwards from Tenby to Whitland to meet the GWR's line from Carmarthen. The opening of the more direct route to South Wales via the Severn Tunnel in 1886 shortened the journey from Paddington by 24¾ miles (see feature box).

The railway west of Swansea assumed a far greater importance once the extension to Fishguard Harbour had opened in 1906. The Swansea District Lines opened six years later thus allowing Ocean Liner Specials, Irish boat trains and heavy cattle trains to and from Fishguard to avoid the choke point of Swansea and the notorious Cockett Bank gradient. With the exception of the branch line to Neyland the entire route is still open for business today, with the section from the Severn Tunnel to Swansea currently being upgraded as part of the Great Western electrification project.

BELOW: A train crosses Brunel's wooden viaduct that carried the broad gauge South Wales Railway over the River Tawe at Landore.

The Severn Tunnel and the Badminton Cut-Off

In 1872 the GWR's Severn Tunnel Railway Act received Parliamentary authorization. Construction soon started from both ends between Pilning in Gloucestershire and Rogiet in Monmouthshire but progress was extremely slow. With pumping engines in continuous action to keep the tunnel from flooding, a large freshwater spring broke into the tunnel when the two headings had nearly met. 6,000 gallons of water each minute overwhelmed the pumps and totally flooded the unfinished tunnel causing work to be abandoned. It was only after divers isolated the spring and the tunnel was pumped dry that construction work could recommence. The two headings met under the river in 1881 but with ongoing flooding problems it was not until 1885 that the brickwork lining and double track could be completed. In September of that year the Chairman of the GWR, Sir Daniel Gooch, travelled through it in a special train. However, the official opening was delayed for a year due to yet more ingress of water that was only solved with the eventual installation of more powerful Cornish beam pumping engines. It finally opened in December 1886 and is still open today, although electric pumps are still required to pump out around thirty million gallons of water each day. The 4-mile 624-yd Severn Tunnel was the longest railway tunnel in the world when opened, and until the completion of HS1 in 2007 remained the longest in Britain.

The flow of traffic through Bristol increased dramatically following the opening of the Severn Tunnel. Heavy coal trains from South Wales to London now competed for line occupancy with the increasingly popular through services from Paddington to the West of England. The GWR tackled this problem by building a new mainline from Wootton Bassett, west of Swindon, to Patchway, where it joined the line from Bristol to the Severn Tunnel. Known as the Badminton Cut-Off this 29½-mile line greatly relieved the Bristol bottleneck when it was opened throughout in 1903.

ABOVE: The Severn Tunnel was the longest railway tunnel in the world when it opened in 1886, shortening the route between London and South Wales by 24¾ miles.

The Waverley Route

Running between Edinburgh and Carlisle, the Waverley Route had humble beginnings in the Edinburgh & Dalkeith Railway, a horse-drawn 4-ft 6-in.-gauge tramway that opened in 1835 to carry coal from mines south of Edinburgh to the docks at Leith. Bought by the fledgling North British Railway in 1845, it was regauged and reopened with steam power in 1847.

Meanwhile the nominally independent Edinburgh & Hawick Railway had already been authorized in 1845 – in reality it was already owned by the North British, which had set its sights on building an Anglo-Scottish route across the Borders to Carlisle. Initially taking in the route of the old Edinburgh & Dalkeith Railway, the line headed off down the valley of Gala Water in a southeasterly direction to Galashiels and Melrose before striking south across hill country to the important mill town of Hawick.

The double-track line was opened in 1849 despite the requirement to build many bridges along the Gala Valley.

The North British Railway's dream of reaching the London & North Western (LNWR) and Caledonian (CR) railways' stronghold of Carlisle took a further thirteen years to come to fruition but in the meantime a much smaller operation, the 13-mile Carlisle & Silloth Bay Railway & Dock Company, held the key to Carlisle's back door. Opening in 1856 the traffic on the line proved disappointing – the company owned no steam engine, but occasionally borrowed one from a neighbouring company, and passenger services were horse drawn. The impoverished railway struggled on for six more years until an unexpected saviour came to its aid.

At the Edinburgh headquarters of the North British Railway, the directors had been planning their next move to reach northwest England and the Border Union (North British) Railway was authorized in 1859 to extend the NBR's Edinburgh & Hawick Railway to Carlisle. Despite strong protests and blocking tactics at Carlisle from both the LNWR and the CR, the new railway was opened throughout in 1862 across the sparsely populated Borders region – the key to the NBR's success in reaching Carlisle was to lease the Carlisle & Silloth Bay Railway & Dock Company at the same time. Serving nothing more than a handful of villages along its route and lonely Riccarton Junction, the double-track railway was a feat of engineering, with the long Whitrope tunnel and Shankend Viaduct standing today as a lasting memorial to their optimistic Victorian engineers.

The new line was romantically advertised as the 'Waverley Route' by the NBR, its route through the countryside immortalized by Sir Walter Scott in his Waverley Novels. Although initially slow to develop as a major trunk route, once the Midland Railway had reached Carlisle in 1876 via the Settle–Carlisle line it came into its own. Through trains were soon running from London St Pancras to Edinburgh via the Settle–Carlisle and the Waverley routes, but despite its scenic attractions this Anglo-Scottish route could never compete with the faster timings achieved on the rival East and West Coast main lines.

Fast forward to 1963, the year in which the infamous 'Beeching Report' was published. Dr Beeching recommended that the entire line should be closed, arguing that its route was already duplicated by the East and West Coast Main Lines. Closure was finally scheduled for 5 January 1969 following several postponements and the last train to travel over it was the Edinburgh to St Pancras sleeper train – among its passengers David Steel, MP for Roxburgh, Selkirk and Peebles.

Feelings were high among local protestors by the time the train arrived at Newcastleton, where they had blocked the line by locking the level crossing gates shut. Tempers were also high and a number of arrests were made before the train was able to continue, but not completing its journey on the Waverley Route until the following day. Tracklifting operations began almost immediately, although this was suspended for some months while fruitless negotiations were held with the Border Union Railway who wanted to keep the line open. Unsurprisingly this all came to nothing and by June 1972 the Waverley Route had been totally ripped up. Large swathes of the Scottish borders were now without any rail access for the first time in 120 years.

In June 2006 the Scottish Parliament overwhelmingly voted to pass the Waverley Railway (Scotland) Act. Reopening the Waverley Route between Edinburgh, Galashiels and Tweedbank had been given the green light. The contract to rebuild the line was awarded to Network Rail and tracklaying commenced in 2014. By the time this book is published trains should once more be running on the northern section of the Waverley Route.

LEFT: Sir Walter Scott Country – the North British Railway proudly proclaimed its line between Edinburgh and Carlisle as the Waverley Route – the 'direct and picturesque route between England and Scotland'. As seen in this 1907 poster, it was also proud of its trains, which included dining cars, sleeping cars and even lavatory carriages.

BELOW: North British Railway's Class 'J' 4-4-0 locomotive No. 895 approaches Hawick with a local passenger train, c.1910. The Waverley Route closed in 1969 but the northern section between Edinburgh and Tweedbank reopened in September 2015.

Settle–Carlisle Line

Starting life as the Leeds & Bradford Railway, the southern section of the Leeds to Carlisle line opened between Leeds and Shipley in 1846 and was extended by the Leeds & Bradford Extension Railway to Skipton in 1848. Both these companies were promoted by the financier George Hudson and were taken over by the Midland Railway (MR) in 1851. The North Western Railway (NWR) further extended the route in 1849 when it opened between Skipton and Ingleton in 1849. The NWR had planned to extend its line northwards from Ingleton to Low Gill where it would meet the Lancaster & Carlisle Railway (worked by the London & North Western Railway) but work on this was halted. The railway between Ingleton and Low Gill was finally built by the Lancaster & Carlisle Railway and opened in 1861, by which time the MR had taken over working the NWR. Theoretically offering a through route for MR trains to Scotland, the Ingleton line was beset by operating problems, exacerbated initially by the fact that the LNWR and MR stations at Ingleton were a mile apart and separated by a viaduct. Through running was deliberately discouraged by the LNWR so the Midland was forced to reconsider its position. The only answer to this problem was to build its own Anglo-Scottish route and so the Settle–Carlisle Line was born.

Surveying for the new 71¼-mile line from Settle Junction to Carlisle commenced in 1865 and it received Parliamentary approval a year later but a national financial crisis soon put a stop to any progress on this £2.3 million project. The Midland Railway sought to abandon the

BELOW: The wind and the rain is lashing across Batty Moss as Stanier class '8F' 2-8-0 No.48151 gets to grips with its heavy stone train, a Ribblehead sidings to Carlisle ballast train crossing Batty Moss viaduct on Tuesday 19 December 2000. Railtrack chartered LMS Stanier-designed Heavy Freight '8F' 2-8-0 48151 from Carnforth, which took a train of twenty hoppers from Hellifield to Ribblehead for loading, then forwarded them to Carlisle in appalling weather. This trip was to mark the reopening of the Settle to Carlisle railway line after its major refurbishment.

RIGHT: A 1907 poster produced by the Midland Railway shows the Settle–Carlisle route as being the integral part of the 'Most Interesting Route to Scotland'. The train at the bottom is being hauled by 4-4-0 No. 1025 *Cock O' the North*, hence the large black bird depicted at the top.

scheme in 1869 but Parliament refused permission on the grounds that other railway companies would suffer, so the company was forced to commence construction work. The statistics for building this line are truly awesome, especially as it was built through the wild Pennine Hills without the aid of mechanical excavators or access along roads. Using temporary tramways, horse-drawn carts, picks, shovels and wheelbarrows, over 6,000 navvies assisted by the then recently invented dynamite blasted their way across the landscape, building twenty-three viaducts (the longest at Ribblehead has twenty-four arches) and thirteen tunnels (the longest, Blea Moor, is 1½ miles long). Ais Gill, the summit of the line, is 1,169 ft above sea level and this is reached from the south on a 15-mile-long ruling gradient of 1 in 100, which became known as the 'Long Drag' by locomotive crews.

Temporary encampments housed the unruly rabble of the workforce along the line, the largest at Ribblehead housing 2,000 including accompanying wives and children, and had all the amenities of a small town. The weather was often atrocious with gales, snowstorms and heavy rain dampening the navvies' spirits. Scenes reminiscent of the Wild West with drunken behaviour were rife and both the local police and preachers were called in to deal with this problem. Many navvies also died, either from work-related accidents or in drunken brawls, and local cemeteries had to be enlarged to cope with the influx of bodies.

In 1875 the railway was finally opened to goods traffic although passengers had to wait a further year before making the epic journey. In the final analysis the Settle–Carlisle Line was completed three years late and cost 70 per cent more than the original estimate. Carlisle Citadel station was enlarged considerably to cope with the extra traffic generated by the Settle–Carlisle Line, for the first time allowing Midland Railway trains to run directly from London St Pancras to Glasgow St Enoch via the Glasgow & South Western Railway and to Edinburgh via the North British Railway's Waverley Route.

Newly introduced Pullman car trains double-headed by Midland Railway crimson lake-liveried locomotives were soon a regular sight on the line. The Midland Railway became part of the London Midland & Scottish Railway (LMS) following the 'Big Four' Grouping of 1923. The two principle expresses of the day were 'The Waverley' (St Pancras to Edinburgh) and 'The Thames–Clyde Express' (St Pancras to Glasgow St Enoch) – both were introduced by the LMS in 1927 although the former was named the 'Thames–Forth Express' until the Second World War. Avoiding a time-wasting stop to replenish water, troughs were installed at Garsdale. Nevertheless, journey times were slow in comparison to the alternative East Coast and West Coast routes, and these expresses' days were numbered by the British Railways era. 'The Waverley' was withdrawn in 1968 and the 'Thames–Clyde Express' in 1974.

Paralleled by the West Coast Main Line (WCML), frequently blocked by snowfall in winter, expensive to maintain and operate, not serving any major centres of population and with little local traffic, it was hardly surprising that the Settle–Carlisle Line was recommended for closure in the 1963 'Beeching Report'. In the end the line's usefulness as a diversionary route when the WCML was closed probably saved the day although all stations along the line apart from Settle and Appleby were closed in 1970. Closure of the line loomed yet again after another scare in the early 1980s, when Ribblehead Viaduct was deemed unsafe. There were only two return trains each day by this time but after fierce and sustained organized opposition the closure was refused in 1989. Since then there has been a transformation of the Settle–Carlisle Line's fortunes, helped in no small way by the 'Friends of Settle Carlisle Line' who have improved facilities and by the reopening of eight stations. There has also been a major investment in the infrastructure – viaducts have been restored, passenger train services improved, through freight workings reintroduced and steam charter trains make regular year-round appearances.

BELOW: An ex-War Department 2-8-0 approaches Blea Moor signal box with a freight train of empty anhydrite hopper wagons in 1958. The photographer was the famous Bishop of Wakefield, Eric Treacy, who sadly died at Appleby station while waiting to photograph a steam railtour in 1978.

RIGHT: The twilight of steam on the Settle–Carlisle. With Ingleborough in the background, Colin Gifford's atmospheric photograph taken in 1967 shows a northbound freight hauled by a BR Class '9F' 2-10-0 near Ribblehead. Standard-gauge steam haulage on the Settle–Carlisle and across British Railways ended at midnight on 11 August 1968.

Great Eastern Railway

Formed in 1862 the Great Eastern Railway (GER) was an amalgamation of many railway companies in East Anglia, of which the Eastern Counties Railway was by far the main constituent. By future takeovers of more companies the GER eventually owned 1,200 route miles throughout Norfolk, Suffolk, Cambridgeshire and Essex, giving it a virtual monopoly of railway services in this region. The monopoly ended with the formation of the Midland & Great Northern Joint Railway in 1893, which allowed the Midland Railway and Great Northern Railway access to Norfolk's important fishing ports and seaside resorts.

In London the GER used the former Eastern Counties Railway terminus at Bishopsgate until its new Liverpool Street station was opened in 1874. The 10-platform overall-roofed station was designed by GER engineer Edward Wilson but within ten years of opening was working to capacity. The company's Great Eastern Hotel was opened in 1884 and an extension to the station including eight new platforms was added in the early 1890s.

The company's main workshops at Stratford had already been established in the 1840s by its predecessor the Eastern Counties Railway. Although locomotive building commenced here in 1850, the GER greatly expanded the site in the 1870s with nearly 1,000 locomotives being built by the end of the century. The works were also home to a succession of talented locomotive superintendents including Samuel W. Johnson, later to become the Chief Mechanical Engineer of the Midland Railway, T.W. Wordsell, who would later hold the same post at the North Eastern Railway, and James Holden, famous for his unique high-pressure 0-10-0 'Decapod' locomotive of 1903. The adjoining engine sheds at Stratford were amongst the biggest in Britain, housing over 500 locomotives at their peak just after the First World War. From 1882 all GER passenger locomotives were finished in a smart blue livery which, when seen at the head of a train of varnished teak carriages, must have been a stirring sight.

The GER were also part owners in several joint lines including the Tottenham & Hampstead Joint Railway, the Norfolk & Suffolk Joint Railway, the East London Railway and the Great Northern & Great Eastern Joint Railway.

FAR LEFT: The Great Eastern Railway operated an intensive commuter service out of London Liverpool Street station. Here a group of workmen get ready to board a train to Enfield Town on 25 October 1884.

LEFT: The coat of arms of the Great Eastern Railway can be seen today at Liverpool Street terminus in London – the station remained virtually unchanged until the mid-1980s when it was redeveloped.

ABOVE: Great Eastern Railway Class 'P43' 4-2-2 locomotive No. 10 was built at Stratford Works in 1898. Designed by James Holden, the ten locomotives of this class had a short working life and had all been withdrawn by 1910.

Completed in 1882 the latter was by far the most important as it allowed the GER access to the coalfields of the East Midlands and South Yorkshire via links to Sleaford, Lincoln and Doncaster.

While much of its network served scattered rural communities the GER also operated one of the busiest commuter operations in the world. The intensive suburban service operating in and out of Liverpool Street station during the morning and early evening rush-hour periods became known as the 'Jazz' – so named after the yellow and blue stripes on carriages denoting first and third class respectively.

For nearly three decades two GER routes vied for supremacy as the fastest mainline service between Liverpool Street and Norwich but by the 1890s the route via Ipswich had won against the alternative via Cambridge. During summer months the GER's lines to the Norfolk seaside resorts became very busy with overflowing trains carrying holidaying Londoners to the delights of Hunstanton, Sheringham, Cromer, Mundesley and Yarmouth.

Freight operations were largely agricultural, along with vast amounts of fish from the ports of Great Yarmouth and Lowestoft destined for Billingsgate Market in London. The GER also served other ports including King's Lynn, Felixstowe and Parkeston Quay with freight trains transporting imported timber for nationwide distribution. Starting in 1863 the company also operated a large number of passenger ferries for North Sea services from the port of Harwich to Rotterdam and Antwerp, while a service from Parkeston Quay to Hook of Holland commenced in 1904. To connect with these the GER introduced fast and luxurious boat trains in 1882 running between Liverpool Street and Harwich, and later Parkeston Quay.

The Great Eastern Railway was one of the major constituent companies that formed the London & North Eastern Railway (LNER) in the 'Big Four' Grouping of 1923.

Southern England

The latter half of the nineteenth century saw rapid expansion of railways across Southern England. The amalgamation and absorption of numerous railway companies resulted in just four large entities – two of which worked together as the South Eastern & Chatham Railway – becoming the constituents of the Southern Railway at the time of the 'Big Four' Grouping.

London, Brighton & South Coast Railway (LB&SCR)

Opening throughout between Norwood Junction and Brighton in 1841 the London & Brighton Railway (L&BR) featured five tunnels through the North and South Downs and the 1,475-ft Balcombe viaduct across the Ouse Valley. At Norwood Junction the L&BR joined the London & Croydon Railway (L&CR) to access London Bridge station. The company shared locomotives with the L&CR and the South Eastern Railway, the arrangement ceasing in 1845. A year later the L&BR amalgamated with the L&CR, Croydon & Epsom Railway, Brighton, Lewes & Hastings Railway and the Brighton & Chichester Railway to form the London, Brighton & South Coast Railway, giving it a network of 170 route miles.

Up until 1860 the LB&SCR used London Bridge station as its London terminus but in that year the 'Brighton' side of Victoria station was opened by the Victoria Station & Pimlico Railway – a consortium of four strange bedfellows in the shape of the Great Western Railway, London & North Western Railway, East Kent Railway and the LB&SCR. By 1865 the LB&SCR had expanded by opening a further 177 route miles of railways in South London, Sussex and Surrey but faced bankruptcy following the banking collapse of 1867. Unlike the London, Chatham & Dover Railway it recovered and went on to expand its South London suburban services, by the late 1880s operating the largest such network in Britain. The pioneering LB&SCR introduced the first British all-Pullman train in 1881 – the 'Pullman Limited Express' between London Victoria and Brighton, which featured the first electrically lit carriages in the country. Suburban electric trains were first introduced in 1909 and by 1921 the company operated 24½ miles of the 'Elevated Electric' out of London Victoria. By the early twentieth century the LB&SCR was in good shape, fairly free from competition, offering its shareholders a good investment return.

South Eastern Railway (SER)

Incorporated in 1836, the South Eastern Railway completed its mainline between Redhill and Dover in 1844 – to reach Redhill from London Bridge station SER trains used the London & Greenwich, London & Croydon and London &

Brighton railway's lines. Before the mainline was completed the SER decided, along with the London & Croydon Railway, to build a terminus in London and in 1844 the terminus at Bricklayers' Arms was opened. It closed again in 1852, when the SER resumed using London Bridge station after leasing the ailing London & Greenwich Railway. In 1846 a line was opened from Ashford to Ramsgate, the SER's locomotive works being established a year later at the former junction. This period saw the opening of the branch line to Folkestone Harbour, the takeover of the Canterbury & Whitstable Railway, the opening of the Medway Valley line from Paddock Wood to Maidstone, and new lines to Gravesend and Strood in North Kent. By 1848 the company was also operating cross-Channel steam-ship services from Dover to Calais, Ostend and Boulogne. A new mainline to Tunbridge Wells and Hastings was completed in 1852, with the Reading, Guildford & Reigate Railway being absorbed in the same year. In 1864 London Bridge was converted into a through station for SER trains to reach Charing Cross station via Hungerford Bridge and two years later the company opened its new City station at Cannon Street.

The latter half of the nineteenth century brought mixed fortunes for the SER with a combination of poor management, disputes with its neighbours and cut-throat competition from its near-bankrupt rival northern neighbour, the London, Chatham & Dover Railway. This situation finally culminated in 1899 with the two companies reaching an agreement to work together, and so the South Eastern & Chatham Railway (SE&CR) was formed – it was not a complete amalgamation as each company still retained its own board of directors. The new SE&CR thrived and with the exception of the independent Kent & East Sussex Railway and East Kent Light Railway it not only controlled the entire rail network in Kent and the important cross-Channel continental traffic but also the Hastings mainline and the important cross-country route across Surrey between Reading, Redhill and Tonbridge.

LEFT: A view of the London, Brighton & South Coast Railway's side of Victoria station in 1887 with one of Wiliam Stroudley's Class 'A1' 0-6-0Ts shunting some ancient looking coaching stock in the middle road. On the left of it a train awaits to depart for Willesden.

BELOW: Opened in 1866, Cannon Street station was the London terminus of the South Eastern Railway. In the foreground is Alexandra Bridge, which was designed by Sir John Wolfe Barry.

London, Chatham & Dover Railway (LC&DR)

The LC&DR's predecessor only by name, the East Kent Railway was in dire financial straits even before it opened its first line between Strood, Chatham and Faversham, in 1858. A year later it changed its name to the LC&DR and by 1865 had extended its route from Faversham to Canterbury and Dover, and also along the north Kent coast to Margate and westwards to London's Victoria station via Swanley and Blackfriars Bridge via Herne Hill and the City branch.

Despite its poor record for punctuality and passenger comfort the LC&DR had a more redeemable quality as being one of the safest railways in Britain, being an early user of the Westinghouse air brake system and fail-safe signalling. However, the banking collapse of 1867 coupled with shady financial dealings between the railway and its contractors, Peto & Betts, led to its downfall and bankruptcy in 1867.

The refinanced LC&DR continued to struggle on through the rest of the nineteenth century, many of its routes having destinations served by its rival, the South Eastern Railway. Sense prevailed and in 1899 the LC&DR entered into a joint working relationship, the new company becoming known as the South Eastern & Chatham Railway.

London & South Western Railway (LSWR)

As we have already seen, the LSWR was born out of the London & Southampton Railway, which had opened throughout from Nine Elms in London to Southampton in 1840. Built on marshland next to the River Thames, Waterloo station replaced Nine Elms in 1848.

By the 1860s the LSWR had expanded its commuter routes from Waterloo out into the leafy suburbs of southwest London, Surrey and east Hampshire but the City of London remained inaccessible until the LSWR-

backed Waterloo & City underground electric railway was opened in 1898. The Portsmouth Direct Line was extended from the LSWR's station at Godalming to Havant via Petersfield by the Portsmouth Railway and taken over following completion by the LSWR in 1859.

Meanwhile the LSWR had its eye on reaching Exeter. With this in mind the meandering Southampton & Dorchester Railway, often referred to as 'Castleman's Corkscrew' after a Wimborne solicitor who had promoted it, opened in 1847 before being amalgamated a year later with the LSWR. From Dorchester LSWR trains later reached Weymouth on mixed-gauge track along the Great Western Railway's broad-gauge line from Westbury and Yeovil that opened in 1857. 'Castleman's Corkscrew' bypassed Bournemouth and the fast-expanding resort was initially reached along a circuitous route from Ringwood in 1870, not being served by the mainline from Brockenhurst until 1888.

The LSWR's ambition to reach Exeter from Dorchester was never realized and instead it opened a new route from Bishopstoke (later renamed Eastleigh) to Salisbury (Milford) via Romsey in 1847. Reaching Salisbury from Basingstoke along what became the company's West of England Main Line took much longer, opening to Andover in 1854 and to Salisbury (Milford) in 1857.

LEFT: Seen here in 1850, a South Eastern Railway passenger train approaches a series of tunnels as it travels on raised tracks at Shakespeare Cliff between Dover and Folkestone.

ABOVE: Designed by Dugald Drummond for the London & South Western Railway and built at Nine Elms Works in 1899, the 'T9' Class 4-4-0s were known for their turn of speed and nicknamed 'Greyhounds'.

Further westward expansion from Salisbury was speedy and in 1859 a through station was opened to coincide with the new extension to Gillingham. Yeovil Junction and Exeter (Queen Street) were reached in 1860 – the latter was linked to the GWR's St David's station via a short steeply graded connection.

The LSWR opened numerous secondary and seaside resort branch lines in the late nineteenth and early twentieth century. Among the last to be opened in 1903 were the Meon Valley Railway between Alton and Fareham and the Lyme Regis branch.

With the Midland Railway, the LSWR also jointly took over the ailing Somerset & Dorset Railway in 1891, the 114-mile route between Bournemouth West and Bath Green Park (along with a branch to Highbridge), giving access to the Midlands and North of England.

Into Cornwall

Cornwall had witnessed a network of horse-drawn tramways and mineral railways from the early years of the nineteenth century despite its isolation from Britain's growing rail network. Built to link the tin and copper mining, china clay extraction and granite quarrying industries with ports and harbours, many of them went on to form the nucleus of the county's modern rail system. One of these, the Bodmin & Wadebridge Railway, was amongst the earliest steam-operated lines in Britain.

Later forming part of the mainline between Plymouth and Penzance, the Hayle & Portreath Railway (H&PR) opened in 1837 to link copper mines around Redruth with harbours at Portreath and Hayle. With inclined planes at Penponds and Angarrack, progress was inevitably slow along this route. In 1852 this all changed when Brunel's West Cornwall Railway (WCR) opened between Truro, Redruth and Penzance. This 25¾-mile railway, originally built to the standard gauge, incorporated part of the route of the H&PR between Redruth and Hayle, bypassing the inclined planes with viaducts.

The broad-gauge double-track Cornwall Railway (CR), also engineered by Brunel, bridged the 53¾-mile gap between Plymouth and Truro and continued another 11¾ miles to the port of Falmouth. The building of this heavily graded line across Cornwall's deep river valleys included numerous high embankments, tunnels and forty-two viaducts, and was a feat of Victorian engineering. The viaducts were originally constructed with timber deck spans and bracing supported on masonry piers – the timber had all been replaced with masonry by the 1930s. The crossing of the River Tamar near Plymouth was achieved by Brunel's *pièce de résistance*, the wrought-iron single-track Royal Albert Bridge, which finally linked Cornwall with the rest of Britain's growing rail network when it opened in 1859.

Despite the opening of the CR, both passengers and freight still had to change trains at Truro, where there was a change of gauge, in order to continue their journey on the WCR to Penzance. When the WCR was relaid to broad gauge in 1866 this operating problem was finally resolved. Although the CR was taken over by the Great Western Railway (GWR) in 1887 and the WCR remained independent until 1948, both lines were operated from their early years by the GWR. Brunel's broad gauge lasted until 1892 when all of the remaining broad-gauge lines west of Exeter were converted to standard gauge over just one weekend. When a new deviation route was opened to the north in 1908, five viaducts between Saltash and St Germans were demolished.

The coming of the main line into Cornwall, along with the opening of numerous branch lines serving harbour towns and seaside resorts, had a significant impact on the county's economy – fishermen, farmers and flower growers could now despatch their fresh produce on overnight trains to London. But by far the biggest impact was the development of many seaside villages and towns into thriving resorts. With holidaymakers now able to access Cornwall's scenic coastline and pristine beaches, tourism had become a major industry by the early twentieth century.

LEFT: Completed in 1859, the Gover Viaduct carried the Cornwall Railway across the Gover Valley near St Austell. Brunel's timber viaducts in Cornwall and Devon were supported on masonry piers and were cheaper to build but they had a shorter life. All were eventually replaced by masonry viaducts built alongside, with that at Gover being replaced in 1898.

ABOVE: The Great Western Railway was undoubtedly the most publicity-conscious railway company in Britain. With St Michael's Mount setting the tone, this 1897 poster promotes tourist tickets and weekly excursions on express and corridor trains to Cornwall. Cornwall was dubbed the 'Cornish Riviera' by the GWR, which also gave the same name to its premier express train.

North Devon and North Cornwall

BARNSTAPLE

The broad-gauge Bristol & Exeter Railway had reached Exeter as early as 1844 but it took another sixteen years for the rival London & South Western Railway (LSWR) to reach the city with the opening throughout of its line from Waterloo via Salisbury and Yeovil Junction. In between these two dates, funded mainly by the LSWR, the Exeter & Crediton Railway had opened as a double-track broad-gauge line between those two places in 1851. It was initially leased to the Bristol & Exeter but finally became part of the LSWR in 1879, by which time the line had already been converted to standard gauge.

Some years earlier in North Devon the Taw Vale Railway & Dock Company had already opened its short route between Fremington Quay and Barnstaple. In 1851 the railway became known as the North Devon Railway & Dock Company and in 1854 it opened a broad-gauge line between Barnstaple and Crediton, thus completing the link with Exeter. The LSWR again had a hand in all this and in 1865 absorbed the company, converting the line to standard gauge in 1876. Meanwhile in 1855 the Bideford Extension Railway had already opened from Fremington to Bideford and was also taken over by the LSWR in 1865. In 1872 a further extension from Bideford to Torrington was opened. North of Barnstaple, the LSWR opened the steeply graded line to Ilfracombe in 1874.

Meanwhile in 1871 the Okehampton Railway opened throughout from Coleford (on the North Devon Railway) to Okehampton in what was to eventually become part of the LSWR's growing empire west of Exeter. With the completion of the 120-ft-high Meldon Viaduct three years later it was extended to Lydford where it met the broad-gauge Launceston & South Devon Railway (an extension of the South Devon & Tavistock Railway). Financially backed by the LSWR, the Okehampton Railway had already changed its name in 1870 to the Devon & Cornwall Railway. In 1872 the LSWR acquired the company and in 1879 opened a branch from Okehampton to Holsworthy, later extending it over Holsworthy Viaduct in 1898 to the small seaside resort of Bude.

LSWR trains initially ran through to Plymouth via Lydford and then along the mixed-gauge track of the South Devon & Tavistock Railway's line through to Marsh Mills Junction. The first trains along this route arrived in Plymouth in 1876 with the LSWR terminus at Devonport being reached over GWR metals through the city. This arrangement was unsatisfactory for the LSWR so in 1883 the authorization of the independent standard-gauge Plymouth, Devonport & South Western Junction Railway between Lydford and the LSWR terminus at Devonport was a joyful occasion! Involving the construction of many bridges and tunnels, the 22½-mile line was expensive to build but opened in 1890 with operations in the hands of the LSWR from the onset. The PD&SWJR remained fairly prosperous and independent until being absorbed by the LSWR in 1922.

To the north, the LSWR's trek westwards to north Cornwall had been painfully slow. The North Cornwall Railway was a subsidiary company and was the next link in the chain, opening from Halwill Junction, on the LSWR's Holsworthy branch, to Launceston in 1886, to Tremeer in 1892, to Camelford and Delabole in 1893 and to Wadebridge in 1895. The final section of what became known as 'The Withered Arm' reached Padstow in 1899, 259¾ miles from London Waterloo.

The completion of this long and tortuous route, and its various tentacles, soon brought great benefits to the widely dispersed rural and coastal communities of North Devon and North Cornwall. Ilfracombe, Bude and Padstow, originally just small fishing villages, all developed rapidly as popular holiday destinations after the arrival of the railway. Farm produce and cattle could reach more lucrative and distant markets than ever before. Vast quantities of milk were collected from country stations along the route and sent to a new milk bottling plant at Vauxhall in south London, fresh fish could be sent by special train from Padstow to the London markets and so could vast quantities of slate from the enormous quarry at Delabole. The LSWR's 'Ocean Liner' specials also competed for traffic between Plymouth and London until the GWR opened its shorter and quicker route via Castle Cary and Westbury in 1906.

LEFT: A nineteenth-century view of Barnstaple with the town's first station in the foreground – the railway link from Exeter was completed in 1854 and taken over by the LSWR in 1865.

ABOVE: Seen here at Little Petherick Creek, the railway to Padstow in North Cornwall was opened by the LSWR in 1899.

Completing the East Coast Main Line

The Scottish county of Fife possessed its own network of railways long before the building of the Tay and Forth bridges, many of them serving the coalfields in the south of the county and later the small fishing villages along the east coast. The main north-south railway linking the Firths of Tay and Forth was the Edinburgh, Perth & Dundee Railway, which was formed by an amalgamation of the Edinburgh, Leith & Granton Railway and the Edinburgh & Northern Railways in 1847. Opening in 1848, this railway linked the Firth of Forth at Burntisland and the Firth of Tay at Tayport. Passengers were carried across the two wide firths by company-owned ferries – five miles across the Forth between Granton and Burntisland and two miles across the Tay from Tayport to Broughty Ferry. From here the Dundee & Arbroath Railway (worked by the Caledonian Railway until 1880) took over for the last few miles into Dundee. Goods were carried separately in the world's first railway wagon ferries, which started operation in 1849.

In 1862 the Edinburgh, Perth & Dundee Railway (EP&DR) was taken over by the North British Railway but passengers between Edinburgh and Dundee continued to endure two uncomfortable ferry crossings or had to take the Caledonian Railway's longer route via Stirling. These working arrangements were becoming totally unsatisfactory although matters improved slightly in 1867 when the NBR introduced a much shorter crossing of the Firth of Forth between North and South Queensferry.

A former manager of the EP&DR, Thomas Bouch had gone on to become a railway engineer, putting forward schemes to build bridges across the two firths as early as 1854. None of these came to fruition not only because of the high cost but also because of the seemingly insoluble engineering problems that would be encountered in building such immense structures. Despite this, Bouch's scheme for a bridge across the Firth of Tay was eventually

BELOW: In 1879 the Tay Bridge collapsed as a train was crossing, killing seventy-five people. Its designer, Sir Thomas Bouch, was disgraced and died a year later.

RIGHT: Produced by the Great Northern Railway, the North Eastern Railway and the North British Railway in 1900, this poster promoted the East Coast route between London and Aberdeen.

STIRLING.
EDINBURGH.
ENGLAND & SCOTLAND
EAST COAST ROUTE
GREAT NORTHERN, NORTH EASTERN & NORTH BRITISH RAILWAYS
FORT WILLIAM.
LOCH LOMOND.
DURHAM
SHORTEST & QUICKEST ROUTE
BETWEEN
LONDON (KING'S CROSS)
AND
EDINBURGH, PERTH, INVERNESS, ABERDEEN &c.
DIRECT ROUTE TO GLASGOW AND WESTERN HIGHLANDS.
DINING CARS
SLEEPING CARS
CORRIDOR TRAINS
AND
LAVATORY CARRIAGES.
FORTH BRIDGE.
YORK.
SCARBOROUGH
McCORQUODALE & Co LIMITED LEEDS.
CHARLES STEEL, General Manager, G.N.R.
GEORGE S. GIBB, General Manager, N.E.R.
W. F. JACKSON, General Manager, N.B.R.

accepted at the end of 1869. Plagued from the outset by design faults and problems with contractors the single-track 85-span cast-iron bridge took nine years to complete and was officially opened by Queen Victoria on 1 June 1878, allowing the NBR to run through trains between North Queensferry and Aberdeen, soon totally eclipsing the Caledonian Railway in passenger numbers and freight tonnage along the route.

On 28 December 1879 one of the worst disasters to befall a railway occurred when North Sea gale force winds brought about the collapse of the centre section of the Tay Bridge's high girders as a train from North Queensferry to Dundee was passing over. All seventy-five passengers, train crew and guard perished in the icy waters of the Tay. The blame for this disaster fell on the bridge's designer and engineer, Thomas Bouch, who had only recently been knighted – he died a broken man a year later.

Following the collapse of the Tay Bridge, Bouch's design for a suspension bridge across the Firth of Forth was scrapped and replaced by a cantilever bridge designed by John Fowler and Benjamin Baker. The main contractor was William Arrol, who also built the replacement double-track Tay Bridge and the Tower Bridge in London. Completed in 1890 the Forth Bridge was the engineering wonder of the world – at 1½ miles long it was by far the largest cantilever bridge in the world and was Britain's first major structure to be built of steel (65,000 tons). This iconic structure has stood the test of time and weather, still performing its task after more than 120 years of continuous use.

In 1887 the new 2¼-mile Tay Bridge had already opened and, with the opening of the Forth Bridge, the NBR finally had its direct route between Edinburgh and Aberdeen. NBR trains North of Dundee ran over the Dundee & Arbroath Joint Railway to Arbroath, originally being opened in 1839 and laid to a 5-ft 6-in. gauge. Converted to standard gauge in 1847, it became a joint undertaking worked by the CR and the NBR in 1880. North of Arbroath the NBR in 1881 opened its direct line through Montrose to Kinnaber Junction – from here NBR trains ran over CR metals to Aberdeen.

The important East Coast route from Edinburgh to Aberdeen was in place in 1890 and through trains were introduced between London King's Cross and Aberdeen, firstly to Doncaster over the GNR, then to Newcastle and Berwick over the NER and finally over the NBR to Edinburgh and Aberdeen. Through trains were being run from King's Cross to Edinburgh as early as 1860 using the specially designed East Coast Joint rolling stock.

The new direct route north of Edinburgh saw a massive increase in freight and passenger traffic for the NBR. In addition to the through expresses from Aberdeen to the south, the route became an important artery for Fifeshire coal, cattle, agricultural produce and fish – all now able to reach important markets in the south overnight. The NBR's locomotive, carriage and wagon works at Glasgow Cowlairs turned out prodigious numbers of successful steam locomotives and rolling stock during the reigns of chief mechanical engineers Matthew Holmes and William Reid, some of the locomotives remaining in service into the 1960s.

BELOW: Although the Forth Bridge opened in 1890, the vehicle ferry that sailed between South and North Queensferry continued to operate until the Forth Road Bridge opened in 1964.

'Railway Races to the North'

At the end of the nineteenth century Kinnaber Junction, north of Montrose, was the finishing post for the famous 'Railway Races to the North'. There was great competition in 1895 between the two rival railway routes for the fastest overnight journey times between London and Scotland. The West Coast Main Line consisted of the LNWR's route from Euston to Carlisle and the CRs route from Carlisle to Aberdeen. The East and West Coast Main Lines from London converged at Kinnaber Junction, both companies striving to beat the other to this point – whoever got there first had won because both trains used the route north of Kinnaber. On 20 August the East Coast Main Line achieved a milestone when the 523-mile journey from King's Cross to Aberdeen was covered in 8 hours 40 minutes at an average speed of just over 60 mph. Although a great publicity coup for the railways involved, passengers were no better off, arriving in Aberdeen at some unearthly hour in the morning. The racing competition came to an end in 1896 when a high-speed derailment occurred at Preston – strict speed limits were then imposed, remaining in force until the early 1930s.

ABOVE: A Caledonian Railway driver inspects his 4-4-0 locomotive No. 17, which was built at St Rollox Works in Glasgow. Designed by J. F. McIntosh, these surefooted 'Dunalastair' engines regularly hauled West Coast Main Line express trains between Carlisle, Glasgow and Aberdeen.

Railway expansion in Scotland

The latter half of the nineteenth century saw a rapid expansion of railway building in Scotland although it took until the beginning of the twentieth century before today's network as we know it was in place.

Caledonian Railway

The Caledonian Railway (CR) opened its main line up the Clyde Valley from Glasgow and Edinburgh to Carlisle in 1849. Following amalgamations and takeovers the 'Caley', as it was affectionately known, ultimately controlled a 2,500-mile network from Wemyss Bay in the west to Aberdeen in the east. Clyde bank commuter lines served Gourock and Dumbarton – on the Ayrshire coast it reached Ardrossan. It stretched across the industrial Central Belt to Edinburgh Princes Street station and deep into NBR territory with branches to Leith and Granton docks.

The route to Stirling and Perth branched off at Dunblane, reaching Oban via a highly scenic route through Callander, Glen Ogle, Crianlarich and the Pass of Brander in 1880. Branches were also opened to Ballachulish and Killin.

Perth was an important junction for the railway with lines radiating out to Glasgow, Dundee, Crieff and Balquhidder via the Almond Valley, and Aberdeen via Stanley Junction, Forfar and Kinnaber Junction – the finishing point of the famous Railway Race to the North. Branching off the Caley mainline in Angus, small towns such as Kirriemuir, Edzell and Brechin were served.

The CR also had running rights over the Glasgow & South Western Railway's line between Dumfries and Castle Douglas and westwards to Stranraer Harbour over the Portpatrick & Wigtownshire Joint Railway. Dumfries was reached on a CR-owned branch line from Lockerbie, on its Clyde Valley route, allowing the company to run its own Irish boat trains.

The 1856 locomotive works at Glasgow St Rollox spawned several eminent Chief Mechanical Engineers including Dugald Drummond, J. F. McIntosh and William Pickersgill. The railway also owned the Caledonian Steam Packet Company, operating Clyde ferry services from 1889. Notable Caley stations included Glasgow Central, and Buchanan Street, Edinburgh Princes Street and Perth. The Caley became a constituent company of the newly formed LMS in 1923.

Highland Railway

The Highland Railway (HR) was formed in 1865 by the amalgamation of the Inverness & Aberdeen Junction Railway and the Inverness & Perth Junction Railway, controlling 242 route miles of railway. The 164½-mile Far North line to Wick and Thurso was completed in 1871 followed in 1884 by acquisitions of the Duke of Sutherland's Railway and the Sutherland & Caithness Railway.

The Dingwall & Skye Railway opened westwards to Stromeferry in 1870 but the final section to Kyle of Lochalsh had to wait for the HR to complete it, opening in 1897. Meanwhile the HR had already absorbed the D&SR in 1880.

The opening of a more direct route between Inverness and Aviemore in 1898 shortened the Highland Main Line. This steeply graded route features the curving 28-arch 600-yd-long Culloden Moor Viaduct and the 445-yd-long steel Findhorn Viaduct.

Company's headquarters and Lochgorm locomotive works were in Inverness, the first locomotive superintendent being none other than William Stroudley who later joined the London, Brighton & South Coast Railway. His successor was David Jones, whose 'Jones Goods Class' was the first 4-6-0 to operate in Britain. At the 'Big Four' Grouping in 1923 the HR and its 494 route miles became part of the newly formed LMS.

LEFT: Designed by J.F. McIntosh and built at St Rollox Works in 1906, Caledonian Railway 4-6-0 No. 903 *Cardean* waits to depart from Glasgow Central station with a West Coast Main Line Express for London Euston in 1908.

BELOW: With snow drifts as high as forty-five feet, Highland Railway locomotives battle to clear the Far North line at Altnabreac on 8 March 1895.

North British Railway

The North British Railway opened the 57-mile railway between Edinburgh and Berwick-upon-Tweed in 1846, today's East Coast Main Line. It also took over the horse-drawn 4-ft 6-in.-gauge Edinburgh & Dalkeith Railway in 1845, converting it to standard gauge and introducing steam power. The NBR expanded rapidly with acquisitions and takeovers. In 1865 it took over the Edinburgh & Glasgow Railway and by the end of the century it had become the largest railway company in Scotland, stretching south to Carlisle via the Waverley Route, north and west to Fort William via the West Highland Line and as far north and east as Kinnaber Junction on the route to Aberdeen. Its densest networks were in the industrial Central Belt and Fife areas. Following the Tay Bridge disaster of 1879 the NBR opened a new bridge over the Firth of Tay in 1887 – it also had a 35 per cent stake in the Forth Bridge, which opened in 1890.

The NBR also backed the building of the West Highland Railway, opening between Glasgow and Fort William in 1894. A triumph of late Victorian engineering, its 198-mile route clings to mountainsides, skirts lochs and crosses miles of featureless and inhospitable bogs. Built by Robert McAlpine ('Concrete Bob'), the 38¼-mile Mallaig Extension from Fort William opened in 1901, showcasing the use of McAlpine's concrete – this highly scenic line features the famous 21-arch Glenfinnan Viaduct. Both the WHR and the Mallaig extension were absorbed by the NBR in 1908.

The NBR operated branch lines in Northumberland and the Carlisle-Silloth line in Cumberland. From small beginnings in 1844 the NBR network had grown to 2,739 miles by the time of the 'Big Four' Grouping in 1923 when it became a major constituent company of the LNER.

Great North of Scotland Railway

In 1854 the Great North of Scotland Railway (GNoSR) opened its first line from Kittybrewster to Huntly, then in 1856 extended northwards from Huntly to Keith and southwards to Aberdeen (Waterloo), using the new joint station in Aberdeen in 1867.

Following a number of absorptions between 1866 and 1881, the GNoSR owned a network of railways throughout northeast Scotland serving the fishing ports of Peterhead, Fraserburgh and Macduff and the important Speyside whisky industry, and running intensive commuter services on the Deeside line.

The Morayshire coastal route was extended in 1886 cutting journey times for GNoSR trains between Aberdeen and Elgin – the alternative route via Keith and Craigellachie was much slower.

The GNoSR opened the Cruden Bay hotel and golf course at the end of the century, connected by a new branch to Ellon on the Fraserburgh line. The Deeside-Ballater line saw frequent royal visits on their way to Balmoral – the first official Royal Train arrived in 1866. In 1903 a new locomotive works was opened at Inverurie, replacing one at Kittybrewster. From 1904 the GNoSR pioneered motor omnibus services with feeders for its many branch lines.

At its peak the GNoSR operated 333½ route miles and in the 'Big Four' Grouping of 1923 became one of the constituent companies of the newly formed LNER.

Glasgow & South Western Railway

The Glasgow & South Western Railway was formed in 1850 by the amalgamation of the Glasgow, Paisley, Kilmarnock & Ayr and the Glasgow, Dumfries & Carlisle Railways – the former had opened its Ayr line in 1840 while the latter had opened in stages along Nithsdale between 1846 and 1850. It opened its locomotive works at Kilmarnock in 1856 where Patrick Stirling was Locomotive

Superintendent until 1866 before he took on the same job at Doncaster for the GNR.

While a shortage of funds prevented the G&SWR from expanding its own network it financially supported many new schemes with working arrangements. These included the Ayr & Dalmellington, Ayr & Maybole Junction Railway, Maybole & Girvan, Ardrossan, and the Castle Douglas & Dumfries, all of which were absorbed between 1854 and 1871. The G&SWR opened a grand terminus at Glasgow St Enoch in 1876. With the opening of the Settle–Carlisle Line in the same year, the Midland Railway came to an agreement with the G&SWR to operate through trains between London St Pancras and St Enoch, the G&SWR opening a grand hotel adjacent to the new terminus in 1879.

The G&SWR drove deep into CR territory by supporting the Greenock & Ayrshire Railway (G&AR), which opened to Greenock's Albert Harbour in 1869. It absorbed the G&AR in 1872 and a price war broke out for the rival Clyde steamer services and competing lines to Glasgow. The G&SWR improved facilities at Greenock in 1875 when it opened Princes Pier station.

The G&SWR's attempts to reach Stranraer harbour – for Northern Ireland – took many years to come to fruition. Already reaching Girvan in 1860 it was 1887 before the G&SWR-backed Ayrshire & Wigtownshire Railway opened between Girvan and Challoch Junction. Absorbed by the G&SWR in 1892, it was then possible for G&SWR trains to reach Stranraer.

Around the turn of the century the G&SWR opened its Paisley Canal Line and lines to Barrhead, Catrine, Dalry, North Johnstone and the Cairn Valley and Maidens & Dunure Light railways, the latter serving the famous golf links and hotel at Turnberry.

The G&SWR operated a fleet of ferries serving the islands of Arran, Greater Cumbrae and Bute, allowing day trippers to escape on trains and ferries from industrial Glasgow to the Ayrshire coastal resorts and Clyde Estuary islands.

It also had a virtual monopoly transporting Ayrshire coal and serving harbours, towns and resorts along the Firth of Clyde coastline. At its peak it operated over 1,000 miles of railway, becoming an important constituent company of the newly formed LMS in 1923.

LEFT: Advertising the scenic beauty of the Western Highlands, this poster produced by the North British Railway in 1920 also publicizes through tickets and carriages from London King's Cross and St Pancras stations.

BELOW: Opened in 1901 the iconic curving 21-arch viaduct at Glenfinnan was built of concrete and is still used today by trains between Fort William and Mallaig.

Great Central Railway

The Great Central Railway (GCR) was the last main line to be built into London until the opening of the 67-mile Channel Tunnel Link, or HS1, between Folkestone and St Pancras in 2007. The GCR had its roots in a much smaller company in existence since 1847, the Manchester, Sheffield & Lincolnshire Railway (MS&LR). It was formed by the amalgamation of four existing railway companies and the Grimsby Docks Company in order to improve communications across the Pennines between the important northern industrial centres of Manchester and Sheffield, the coalfields of south Yorkshire and the fast-growing port of Grimsby. The MS&LR soon flourished under its ambitious General Manager and later Chairman, Edward Watkin, and expanded its territory through takeovers and the arrangement of joint running rights with other railways in the region. Until then, the MS&LR, a major coal carrier, was a thriving concern despite intense competition from its larger rivals. However, Edward Watkin had a dream of building a railway tunnel under the English Channel and to this end planned further expansion southwards to London.

The new railway, to a new London terminus at Marylebone from Annesley in Nottinghamshire, would transform his company from a regional player into a major north-south strategic route able to compete head on with giants such as the Midland Railway and the Great Northern Railway. One section of the line, southward from Quainton Road in Buckinghamshire to Harrow, was to be jointly run with the Metropolitan Railway, of which Edward Watkin was also Chairman. The Metropolitan had already started operating suburban services on this route in 1892.

Parliamentary approval for the 'London Extension' was obtained by the MS&LR in 1893 and four years later Watkin changed the name of his company to the Great Central Railway. In 1895 work started on the construction of the new 92-mile line, opening for passenger traffic on

15 March 1899. The line was designed from the outset for fast traffic with a continental loading gauge and no level crossings. Major towns and cities that were served by the Great Central included Nottingham, Loughborough, Leicester and Rugby – all of which were already well served by other major railway companies – and the company soon ran into trouble as the actual cost of building the new line far exceeded the original estimate of £6 million.

There was a major falling out between the GCR and the Metropolitan Railway before the new main line had been completed and so an 'Alternative Route' was built from Grendon Underwood Junction on the GCR main line and Ashenden Junction on the Great Western Railway's line from Birmingham and Banbury. From Ashenden Juncton to Northolt Junction the line was jointly owned by the GCR and GWR. The 'Alternative Route' was opened in 1906 by which time the GCR and the Metropolitan had patched up their differences, making the new line rather superfluous.

The most important part of the new Great Central, despite the main line reaching London, was the 'branch' between Culworth Junction, south of Woodford & Hinton (renamed Woodford Halse in 1948), and Banbury, which carried vast quantities of freight and coal traffic and also inter-regional cross-country passenger trains. Woodford, once a sleepy village, soon grew into a major railway junction – this was further enhanced when a new spur was built linking it to another major freight-carrying line, the east-west Stratford-upon-Avon & Midland Junction Railway.

LEFT: A crowd watches the departure of the first train – hauled by Class '11A' 4-4-0 No. 861 – from London Marylebone, the Great Central Railway's new terminus in London, in 1899.

ABOVE: Two Great Central Railway 4-4-0s head into the asphyxiating single-bore Woodhead Tunnel in 1903 – when opened in 1845 the 3-mile 13-yd-long tunnel was the longest railway tunnel in the world.

Alexander Henderson took over as Chairman of the Great Central after Edward Watkin retired in 1899. Sam Fay took over as General Manager in 1902 and, in the same year, John Robinson was appointed Chief Mechanical Engineer. The company headquarters were moved to Marylebone in 1905. In Robinson the Great Central had found a very able locomotive and rolling stock designer and soon the company was living up to its publicity for 'Rapid Travel in Luxury'. Passenger traffic never lived up to expectations despite this, but the railway did achieve much success as a freight carrier, especially excelling in the movement of coal. In the 1923 'Big Four' Grouping the Great Central became part of the newly formed London & North Eastern Railway, and on nationalization in 1948 was allocated to the Eastern Region of British Railways.

The first underground railways

Metropolitan Railway

By the mid-nineteenth century London was becoming a very busy and congested capital. There were already seven railway termini with only one, Fenchurch Street, serving the City itself. Various proposals were put forward to link the termini with the City by the building of an urban railway and in 1854 the Metropolitan Railway ('Met') received Royal Assent. The route of the new railway took it from the Great Western Railway's Paddington station eastwards to the London & North Western Railway's Euston station and the Great Northern Railway's King's Cross station before ending at Farringdon Street in the City.

Most of the 3¾-mile railway was to be underground using the 'cut-and-cover' method – shallow tunnels were first excavated from ground level as a trench and then covered over to allow roads or buildings to be built above. Construction started in 1860 but inevitably there were problems with building subsidence and flooding of the workings, the latter occurring when the underground River Fleet burst through at Farringdon. The double-track railway was initially laid as mixed gauge to accommodate the broad-gauge trains of the GWR and the standard-gauge trains of the GNR.

The 'Met' was opened on 10 January 1863 and was an instant hit with Londoners, carrying nearly 40,000 passengers on its first day and nearly ten million in its first year. Building on this success the 'Met' was extended from Farringdon Street to Moorgate Street, opening in 1865. Allowing through traffic from the London, Chatham & Dover Railway's station at Blackfriars Bridge via the newly opened Snow Hill Tunnel under the Thames, an extra pair of standard-gauge tracks were opened to King's Cross and the site of the Midland Railway's new terminus at St Pancras a year later.

By 1873 the 'Met' had extended westwards from Paddington to Hammersmith and Kensington Addison Road and southwards to South Kensington. With further acquisitions and new construction it had expanded further into the North London suburbs and out into the countryside by the end of the century. From its station at Baker Street its tentacles reached Harrow, Watford, Amersham, Aylesbury, Quainton Road and Verney Junction – services from Baker Street to the latter station in Buckinghamshire commenced in 1897. In 1899 the 'Met' leased the Duke of Buckinghamshire's 6½-mile Brill Tramway from Quainton Road to Brill.

Until electrification in the early twentieth century, all trains on the 'Met' were steam hauled but despite condensing apparatus on the locomotives the atmosphere in the tunnels was particularly unpleasant. Using a third-rail system pioneered by the City & South London Railway in 1890 (see below) the first electric trains started running in 1905 although trains north of Harrow continued to be steam hauled for many more years. With the coming of the Metropolitan Railway an extensive ribbon development of new housing sprang up in leafy Middlesex, Hertfordshire and Buckinghamshire, soon earning it the name of 'Metro-land' – a term dreamt up by the Met's publicity department in 1915.

City & South London Railway 1890

Opened in 1890 the City & South London Railway (C&SLR) was the first deep-level underground tube railway in the world and also the first to use electric traction. The 3¼-mile line ran in a pair of tunnels beneath the Thames between the City of London and Stockwell, constructed using a tunnelling shield and cast-iron segments invented by South-African born James Henry Greathead. Although originally designed to be cable hauled, electric traction was finally chosen, with current collected from a third rail laid beneath the trains. Trains comprised three coaches – later nicknamed 'padded cells' – hauled by a diminutive electric locomotive with power generated at a small power station located at Stockwell. Serving stations at Stockwell, The Oval, Kennington, Elephant & Castle, Borough and King William Street the railway was a great success, carrying five million passengers in its first year of operation.

In 1900 the C&SLR opened extensions northwards to London Bridge, Bank and Moorgate Street and southwards to Clapham North and Clapham Common. A year later the northern section was further extended to Old Street, City Road and Angel and in 1907 to King's Cross and Euston. What is today London Underground's Northern Line was rapidly taking shape.

Glasgow Subway 1896

Glasgow's first underground railway was a 3-mile section of the Glasgow City & District Railway, which opened in 1863. The Glasgow Subway, opened in 1896, is the third-oldest underground metro system in the world. With its headquarters at St Enoch, the Glasgow District Subway

Corporation was unusual as its railway was built to a gauge of 4 ft and operated trains on a 6½-mile double-track circular route powered by a clutch and cable system. The two lines, twice burrowing under the River Clyde, were known as the Outer Circle and the Inner Circle and served fifteen stations. Trains ran clockwise and anticlockwise respectively, taking twenty-four minutes for a complete circuit – each line ran in a separate tunnel and the continuously moving cables were driven from a steam-powered winding station located south of the river near Shields Road station.

Glasgow Corporation took over the subway in 1923, electrifying the lines in 1935 and converting the original 'gripper' carriages to operate on 600V DC collected via a third rail. These late nineteenth-century vehicles, some retaining their original lattice gates, remained in service until 1977 when the whole system was closed down for modernization. Power for train lighting came from conductor rails located on the sides of the tunnels, a system that had been installed earlier when the line was still cable powered. Until modernization in 1977 there was no direct connection with the railway's workshops at Govan so trains requiring servicing had to be lifted off the running lines by a crane.

BELOW: A view of the 'Widened Lines' near Moorgate that opened in 1866 – at the bottom is the new Great Northern Railway line while above it a Metropolitan Railway train heads for King's Cross.

Central and West Wales

Mainly concentrated in South Wales, horse-drawn wagonways had already become an important form of transport in Wales during the early years of the Industrial Revolution in the eighteenth and early nineteenth centuries. One of these, the Penydarren Plateway near Merthyr Tydfil, became the first railway in the world to experiment with steam haulage in 1804. During the early years of the Railway Revolution the two most important railways in Wales to connect with Britain's growing network were Robert Stephenson's standard-gauge 1850 Chester & Holyhead Railway and Brunel's broad-gauge 1856 South Wales Railway. To the east, the mainline along the Welsh Marches between Newport, Abergavenny, Hereford, Shrewsbury and Chester had been completed in 1853.

Penetration of remote Central Wales and the slate-shipment harbours of West Wales was initially achieved in a piecemeal fashion by many local railway companies but by the 1870s the main routes of the Cambrian Line and the Central Wales Line were owned by just two – the former by the Cambrian Railways and the latter almost completely by the LNWR.

Cambrian Railways

The Cambrian Railways (CR) was formed in 1864 by the amalgamation of four Welsh railways: the Oswestry Ellesmere & Welshpool, Oswestry & Newtown, Llanidloes & Newtown and Newtown & Machynlleth railways. A year later the Aberystwyth & Welch [sic] Coast Railway was absorbed and the CR opened its works with its headquarters in Oswestry, nearly doubling the town's population within a year.

Following a financial crisis the company went on to build the Wrexham & Ellesmere Railway, which opened in 1895 – a lifting drawbridge section in the 764-yd-long Barmouth Bridge across the Mawddach Estuary was installed in 1899. The coming of the railway to the coastal destinations of Aberystwyth, Barmouth and Pwllheli rapidly led to an influx of tourist traffic from the industrial Midlands and Northwest England during the summer months, providing an important source of income for the CR. In 1903 the company opened the narrow-gauge Welshpool & Llanfair Light Railway, the pattern of absorbing other railways continuing with the

LEFT: After absorbing the Cambrian Railways in 1922 the GWR enthusiastically promoted the magnificent scenery and allure of West Wales and even published a book of 'Rambles around the Cambrian Coast'.

ABOVE: Barmouth Bridge carries the Cambrian Coast line across the Mawddach Estuary near Barmouth. Built mainly of timber, the 764-yd-long bridge was opened in 1867 and has a swing bridge section to allow tall ships to pass.

Mid-Wales in 1904, the narrow-gauge Vale of Rheidol 1913 and the Tanat Valley Light Railway in 1921.

The CR suffered its worst accident on 26 January 1921 when two trains collided head-on at Abermule station killing fifteen passengers. Within a year the CR was absorbed by the GWR. At its peak in 1921 the CR owned 230 route miles of railways stretching from Pwllheli and Aberystwyth in the west, Brecon in the south, Wrexham in the north and Whitchurch in the east. The mainly single-track Cambrian lines from Welshpool (via Shrewsbury) to Aberystwyth and Pwllheli are still open for business today while the company's two former narrow-gauge lines along the Banwy and Rheidol Valleys are operated as heritage railways.

Central Wales Line

Seen by the LNWR as a way of encroaching on GWR territory, the 110-mile Central Wales Line between Craven Arms and Swansea took many years to complete. In the south the Llanelly Railway had already reached Llandovery from Swansea by 1865. To the north the Knighton Railway (KR) had opened between Craven Arms, on the Shrewsbury & Hereford Railway (S&HR), to Knighton in 1861. A year later, the LNWR, the GWR and the West Midland Railway jointly leased the S&HR. With a watchful eye being kept on proceedings from Euston, the KR merged with the uncompleted Central Wales Railway (CWR) in 1863 and opened for business between Knighton and Llandrindod Wells in 1865. In 1868 the Central Wales Extension Railway (CWER) completed the steeply graded missing link, which included the single-bore Sugar Loaf Tunnel and the curving Cynghordy Viaduct, between Llandrindod Wells and Llandovery.

The LNWR moved fast, absorbing both the CWR and the CWER within a few months of the latter's opening. However it took until 1873 before the LNWR was granted running powers over the Llanelly Railway's Swansea route

– the company now had a through route from Euston to Swansea, albeit 278 miles against the GWR's more direct 191-mile route from Paddington.

Soon, through coaches were serving the Central Wales Line not only from Euston but also from York, Birmingham, Liverpool and Manchester. Journey times along the heavily graded single-track route were slow out of necessity, but the small spa resorts of Llandrindod Wells, Builth Wells, Llangammarch Wells and Llanwrtyd Wells all benefitted from an influx of Victorian visitors seeking the health-curing properties of their mineral spring waters. Freight traffic was heavy, with trains of anthracite headed for the North and the Midlands labouring up to the choking confines of Sugar Loaf Tunnel. One through working that survived until the end of steam haulage in 1964 was the weekday overnight mail train between York and Swansea Victoria – also conveying passenger coaches, it took over nine hours to complete the journey.

Listed for closure in the 1963 'Beeching Report' and surviving another threat in 1969, the highly scenic Heart of Wales Line, as it is now known, is fortunately still open for business today.

Other nineteenth-century routes

Prior to its absorption of the Cambrian Railways in 1922, the GWR had already acquired its first route to the West Wales coast. The 54½-mile single-track line from Ruabon to Barmouth was completed in 1870, giving the GWR access to the growing resorts of the West Wales coastline. A 'Dr Beeching victim', this scenic line closed in January 1965 although the Llangollen to Corwen section has since reopened as a heritage railway.

In the south the Brecon & Merthyr Railway had finally reached Brecon from Newport via Merthyr Tydfil and Torpantau in 1868 – closure came at the end of 1962. The Midland Railway also reached into mid and South Wales by way of the Hereford, Hay & Brecon

Railway and the Swansea Vale Railway, both of which it absorbed in 1876. Through trains were introduced between Birmingham and Swansea via Hereford and Brecon a year later, continuing until the end of 1930 – Brecon lost all its railway connections at the end of 1962.

To the southwest the grandly named Manchester & Milford Railway opened between Aberystwyth and Carmarthen in 1867. Never a great success, this rural line through sleepy West Wales was absorbed by the GWR in 1911 – another victim of Dr Beeching, it closed to passengers in February 1965.

In the northwest the LNWR trod on the Cambrian Railways' toes by acquiring the Bangor & Carnarvon Railway in 1857 and the Carnarvonshire Railway in 1870, allowing it to operate through summer holiday services direct from Euston and the Midlands via the North Wales coastal route to the resorts of Porthmadog and Pwllheli. Both victims of Dr Beeching, the line south of Carnaerfon to Afon Wen closed in December 1964 while the section from Bangor to Caernarfon closed in 1970.

LEFT: An Edwardian scene in Central Wales – a London-bound train hauled by two London & North Western Railway 2-4-2T locomotives enters Llandrindod Wells station, c.1905.

BELOW: A poster produced by the GWR in the 1920s promotes rail travel to the Cambrian Coast and the Dee Valley. Apart from the main line between Welshpool and Aberystwyth and the coastal route from Machynlleth to Pwllheli which are still open, all the other routes shown were closed in the 1960s.

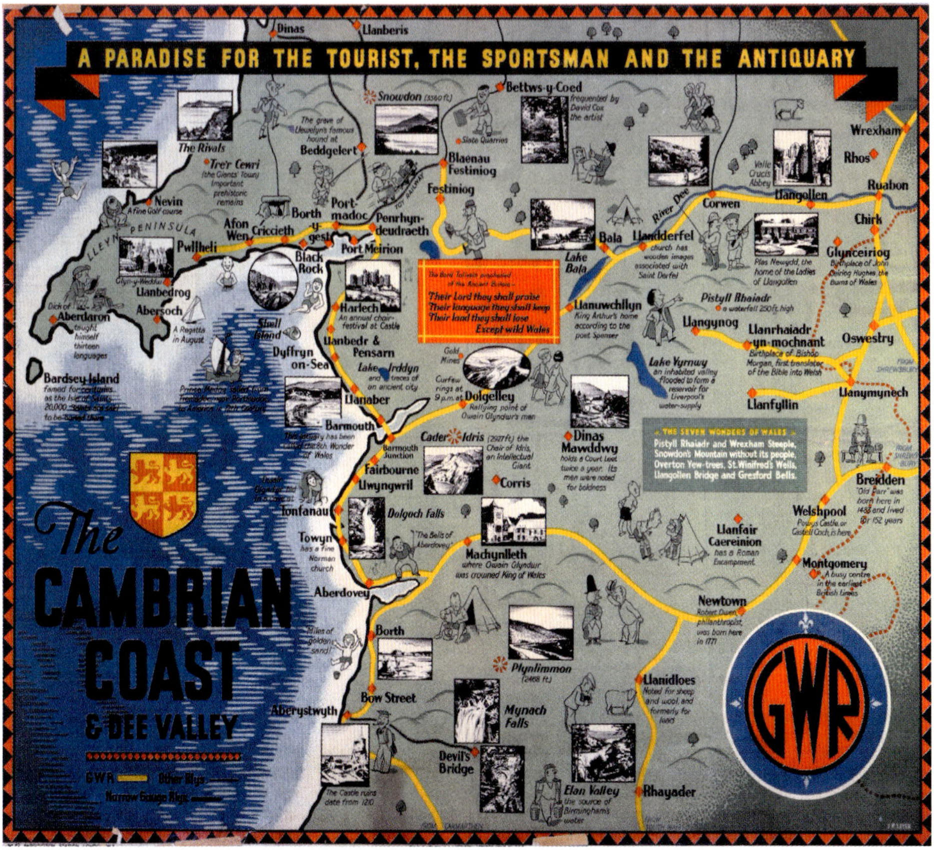

Welsh narrow-gauge railways

With its thriving slate-quarrying industry, harbours and mountainous terrain, North Wales became the centre of narrow-gauge railways in the nineteenth century. Narrow-gauge lines were particularly suited to the mountainous terrain of North Wales, their contour-hugging sharper curves and steep gradients requiring less heavy engineering and being cheaper to build. The earliest lines were simply horse-drawn tramways with loaded slate wagons travelling by gravity from slate quarries to the nearest harbour, the empty wagons being returned to the quarry by horsepower. Virtually no standardization of gauge existed, most of the railways being isolated affairs with few of them linked up – gauges varied from 1 ft 10¾ in. to 4 ft. The pioneering Ffestiniog Railway introduced the first narrow-gauge steam locomotives in 1863 and by the end of the century horsepower was restricted to some of the shorter undertakings and workings in quarries. The vast majority of these Welsh railways were in the north of the principality but a few other isolated examples were subsequently opened elsewhere. Thanks to the pioneering preservationists who saved the Talyllyn Railway from closure in 1951, many have been reborn in more recent years as heritage railways. The following list is not exhaustive but covers all major operators.

Corris Railway
Machynlleth to Aberllefeni and branches to quarries
Gauge: 2 ft 3 in. *Opening date:* 1859
Transported slate from quarries around Corris as a horse tramway.
1879 – Steam locomotives introduced
1883 – Passenger services introduced (ceasing in 1931)
1930 – taken over by the GWR and closing in August 1948 following severe flooding. A ¾-mile section between Corris and Maespoeth has since reopened as a heritage railway

Croesor Tramway
Croesor slate quarries to Porthmadog
Gauge: 2 ft *Opening date:* 1864
This 6½-mile line was operated entirely using horsepower.
1902 – Purchased by Portmadoc, Beddgelert & South Snowdon Railway

BELOW: Built in 1879 at Boston Lodge Works, Double Fairlie 0-4-4-0T No. 10 *Merddin Emrys* hauls a train of slate wagons at Dduallt on the Ffestiniog Railway. After closing in 1946 the railway was gradually reopened between 1955 and 1982.

1922 – Croesor Junction to Portmadog section taken over by Welsh Highland Railway (see below), rebuilt for steam operations. Upper section of line used horsepower until end of the Second World War when it closed

Ffestiniog Railway
Blaenau Ffestiniog to Porthmadog
Gauge: 1 ft 11½ in. *Opening date:* 1836
Originally this 13½-mile line was a horse-drawn and gravity tramway that carried slate from quarries at Blaenau Ffestiniog down to Porthmadog Harbour.
1863 – Steam power in the shape of diminutive 0-4-0s introduced
1865 – Passenger services introduced
1869 – First of several Double Fairlie 0-4-4-0 articulated locomotives introduced
1939 – Passenger services withdrawn
1946 – Closure following rapidly declining slate traffic
1955 – Preservation group reopened the line in stages to 1982

Glyn Valley Tramway
Chirk to Glyn Ceiriog
Gauge: 2 ft 4½ in. *Opening date:* 1873
Opened to serve granite and slate quarries in the Ceiriog Valley, this mainly roadside tramway connected initially with the Shropshire Union Canal at Chirk Bank.
1888 – Rebuilt for steam haulage using Beyer Peacock tram locomotives
1933 – Passengers services withdrawn
1935 – 8¼-mile line closed completely

Nantlle Railway
Nantlle Valley to Caernarfon
Gauge: 3 ft 6 in. *Opening date:* 1828
Built by Robert Stephenson to transport slate from quarries in the Nantlle Valley down to the harbour at Caernarfon, the railway was exclusively operated using horsepower.
1865 – Absorbed by the Carnarvonshire Railway, becoming part of LNWR
1867 – Northern section of the line rebuilt to standard gauge and used by LNWR as its new route to Afon Wen Remainder of line from transshipment sidings at Talysarn to the Nantlle quarries remained horse drawn until 1963 – the last recorded use of horsepower by British Railways!

Padarn Railway
Dinorwic slate quarry to Port Dinorwic
Gauge: 4 ft *Opening date:* 1842
Built to transport slate from the vast Dinorwic Quarry near Llanberis down to the coast at Port Dinorwic. It replaced the Dinorwic Railway, a 2-ft-gauge horse-drawn tramway that opened in 1812. Slate was carried in quarry wagons (with a gauge of 1 ft 10¾ in.) that were loaded on to 4-ft-gauge transporter wagons.
1848 – Steam power introduced – original locomotive *Fire Queen* now on display at the National Trust's Penrhyn Castle Railway Museum
1961 – Railway closed – part of its trackbed now used by the 1-ft 11½-in. Llanberis Lake Railway

Penrhyn Railway
Bethesda slate quarries to Port Penrhyn at Bangor
Gauge: 1 ft 10¾ in. *Opening date:* 1801
Transporting slate from Lord Penrhyn's slate quarries at Bethesda to the coast at Port Penrhyn, this was a horse tramway originally built to 2-ft 0½-in. gauge.
1878 – Rebuilt to 1-ft 10¾-in. gauge using steam locomotives
1962 – Railway abandoned

Snowdon Mountain Railway
Llanberis to Snowdon Summit
Gauge: 2 ft 7½ in. *Opening date:* 1896
Built to carry tourists to the top of Mt Snowdon, 3,493 ft above sea level, this 4½-mile steeply graded railway still operates using the Swiss Abt rack-and-pinion system.
1896 – Fatal accident on opening day closed the line
1897 – Reopened and apart from wartime has operated ever since, still using some of the original steam locomotives, supplemented by modern diesels

Talyllyn Railway
Bryn Eglwys slate quarries to Tywyn Harbour
Gauge: 2 ft 3 in. *Opening date:* 1866
Operated from the outset by steam locomotives, this 7¼-mile railway transported slate from quarries at Bryn Eglwys to the harbour at Tywyn. Passenger traffic, especially tourists in the summer, was also important with trains running between Abergynolwyn and Tywyn.
1911 – Quarries and rundown railway purchased by local MP, Sir Haydn Jones

1920s/30s – Slate industry in decline but tourist traffic kept the line alive
1950 – Closure seemed inevitable upon Sir Haydn Jones' death
1951 – Saved by a group of preservationists, reopening on Whit Monday, the first railway preservation scheme in the world continuing to operate to this day

Vale of Rheidol Railway
Aberystwyth to Devil's Bridge
Gauge: 1 ft 11¾ in. *Opening date:* 1902
Originally built to serve lead mines, forestry operations and tourism.
1913 – Vale of Rheidol Railway (VoR) absorbed by the Cambrian Railways
1920 – Goods traffic ceased
1922 – Became part of the GWR and developed as tourist attraction
1948 – Became part of British Railways – last steam operated on BR
1989 – Sold to private company. Still uses original locomotives on 11¾-mile scenic route

Welsh Highland Railway
Porthmadog Harbour to Dinas Junction/Caernarfon
Gauge: 1 ft 11½ in. *Opening date:* 1922/2011
The Welsh Highland Railway (WHR) was formed in 1922 but its origins date back to 1877 when the North Wales Narrow Gauge Railway opened between Dinas Junction and slate quarries at Bryngwn.
1881 – Extended to South Snowdon (Rhyd Ddu) but there it stopped – the grand dream of a network of narrow-gauge lines in North Wales had been abandoned
1923 – Route between Dinas Junction and Porthmadog eventually completed by the WHR
1934 – Financially was a dismal failure and after struggling on in receivership, leased to the Ffestiniog Railway (FR)
1937 – Despite tourism potential the FR was also unsuccessful and the line closed
1995 – After much legal wrangling, the Ffestiniog Railway was allowed to take on the rebuilding and operation of this highly scenic railway. Aided by major lottery funding and grants the line reopened in stages between 1997 and 2011 – the first three miles between Caernarfon and Dinas were laid on the trackbed of the standard-gauge line that closed in 1964. Today's tourists enjoy this highly scenic 22-mile journey through the Snowdonia National Park in trains hauled by former South African Railways' articulated steam locomotives

Welshpool & Llanfair Light Railway
Welshpool to Llanfair Caereinion
Gauge: 2 ft 6 in. *Opening date:* 1903
Built under the provisions of the 1896 Light Railways Act this rural, steeply graded, 8½-mile steam-operated railway was opened along the Banwy Valley to link farming communities with the market town of Welshpool, operated by the Cambrian Railways.
1922 – Absorbed by the GWR
1931 – GWR ended passenger traffic
1948 – Taken over by British Railways
1956 – Closed by British Railways
1963 – Preservation group stepped in, reopening the line in stages between Llanfair Caereinion and Welshpool (Raven Square) between 1963 and 1981. The two original Beyer Peacock locomotives are still in use today

LEFT: Built by Beyer Peacock in 1902, GWR 0-6-0T No. 822 heads a freight train on the 2-ft 6-in.-gauge Welshpool & Llanfair Light Railway, c.1950. The line was closed by BR in 1956 but has since reopened between Welshpool (Raven Square) and Llanfair Caereinion.

ABOVE LEFT: In 1951 the 2-ft 3-in.-gauge Talyllyn Railway became the first in the world to be saved from closure by preservationists. It continues to operate today using original locomotives and rolling stock.

ABOVE RIGHT: Opened in 1896, the 2-ft 7½-in.-gauge Snowdon Mountain Railway still carries passengers to the 3,493-ft summit of Snowdon using steam and diesel locomotives operating with the Swiss Abt rack-and-pinion system.

The Welsh Valleys

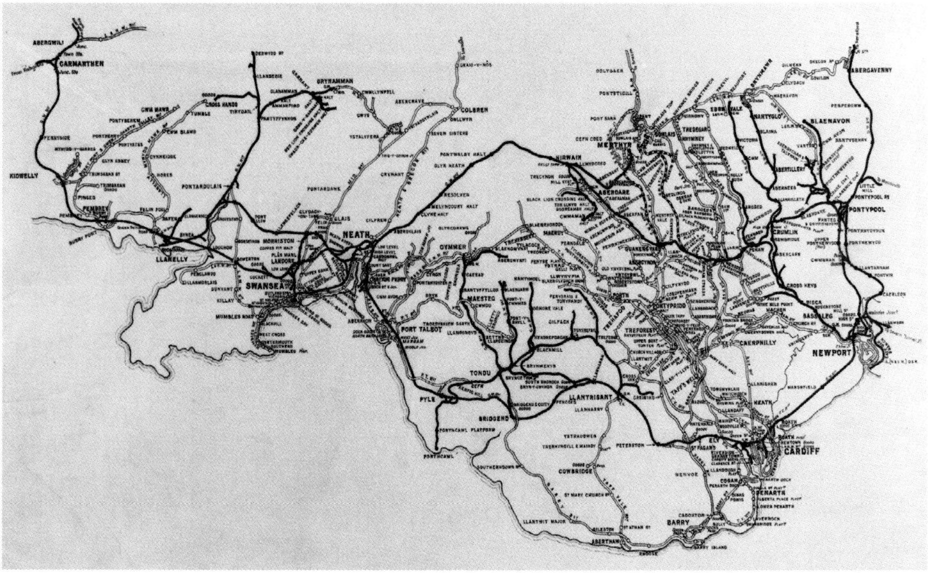

By the end of the nineteenth century the narrow valleys of South Wales contained one of the densest railway networks in the world. This intricate network was made up of numerous independent railway companies all of whom had one common aim – transporting coal from collieries to coastal docks. Unlike the northeast of England where coal transport was in the grip of one company, namely the North Eastern Railway, in South Wales there were sixteen public railway companies in operation. Peaking before the First World War, South Wales was the largest coal-producing region in Britain, annually exporting worldwide over thirty-five million tons of coal from its purpose-built docks.

Barry Dock & Railway Company

Owning sixty-eight railway route miles the Barry Dock & Railway Company was built to break the monopoly of the Taff Vale Railway and Cardiff Docks. Bypassing the TVR's congested line between Pontypridd and Cardiff Docks, the railway ran from Trehafod Junction (for Rhondda Valley coal) and Barry Junction (for Merthyr and Rhymney Valleys coal) to the new Barry docks complex. Opening in 1889, it was engineered by Sir John Wolfe Barry and Henry Marc Brunel, featured two long tunnels and four soaring viaducts – Walnut Tree Viaduct being the most well-known. The company also opened the Vale of Glamorgan Line between Barry and Bridgend in 1897. A highly profitable concern, it was renamed Barry Railway (BR) in 1891 and went on to transport 30 per cent of South Wales' coal for export – by the First World War Barry was the largest coal-exporting port in the world. Primarily a coal-carrying line, the BR also became busy during summer weekends and bank holidays when it transported thousands of day trippers from the valleys to Barry Island. It became a constituent of the GWR in 1922. Apart from a short section in Barry, which still survives, the rest of the BR had closed to all traffic by 1964.

Brecon & Merthyr Tydfil Junction Railway

Opening across the remote Brecon Beacons between Brecon and Merthyr Tydfil in 1868, the Brecon & Merthyr Tydfil Junction Railway went on to purchase the Rumney Railway in 1868, allowing direct access to Newport Docks. While steeply graded northern half of the line via Torpantau Tunnel and Talybont-on-Usk carried little freight, the southern section to Newport Docks was a major coal carrier. The railway became a subsidiary of the GWR in 1922. The end for this 47-mile line between Brecon and Newport came in 1962 when passenger services ceased, although coal traffic continued on the southern section until 1985.

Great Western Railway

South Wales was served by Brunel's South Wales Railway, which had opened between Newport, Cardiff and Swansea in 1850 and amalgamated with the GWR in 1862. Unlike other railways in South Wales it was originally built to broad gauge, only being converted to standard gauge in 1872. Before the 'Big Four' Grouping of 1923 the GWR had already extended its network into South Wales' Valleys by a series of acquisitions and new construction. Notable amongst these was the as-yet unfinished but strategically important line along the heads of the valleys between Pontypool and Neath, which it acquired in 1863. The GWR also went on to own the Newport to Ebbw Vale and Brynmawr routes in the west Monmouthshire valleys and branches radiating out from Bridgend and Tondu to collieries at Abergwynfi, Blaengarw, Nantymoel, Gilfach and Blaenglydach.

ABOVE: Colin Gifford's photograph taken in March 1961 shows a former GWR Class '4200' 2-8-0T climbing Ebbw Vale under Crumlin Viaduct with an empty coal train. The 200-ft-high iron truss viaduct carried the Taff Vale Extension Railway and opened in 1857 – it was not only the highest viaduct in Britain but also the third highest in the world. Sadly it was dismantled in 1967.

LEFT: This 1921 map of the railways in the South Wales Valleys clearly shows the intricate network of lines that once served the valleys, their coalfields and docks.

London & North Western Railway

The LNWR's tentacles stretched far and wide during the latter half of the nineteenth century. By 1873 it had already penetrated as far as Swansea by way of the Central Wales Line but its next move was bolder. By absorbing the unfinished Merthyr, Tredegar & Abergavenny Railway in 1866, it controlled an important route along the heads of the South Wales Valleys giving access to lucrative coal traffic. The entire 24½-mile switchback route was completed between Abergavenny and Merthyr in 1879. A further incursion into 'foreign lands' had occurred in 1876 when the LNWR absorbed the Sirhowy Railway, making Nine Mile Point the company's most southerly incursion into the South Wales Valleys – a sore point not missed by the rival GWR. By gaining running powers over the GWR between Nine Mile Point and Newport, the LNWR was able to operate passenger services along the 24½ miles from Nantybwch on the Heads of the Valleys Line north of Tredegar, down the Sirhowy Valley and into GWR territory at Newport. The LNWR also accessed Cardiff Docks with working arrangements over the Rhymney Railway (see below).

Midland Railway

The Midland Railway also reached into South Wales via the Hereford, Hay & Brecon Railway (absorbed 1876), the Neath & Brecon Railway (running powers) and the Swansea Vale Railway (absorbed 1876). Through trains were introduced between Birmingham and Swansea via Hereford and Brecon a year later. By taking over the SVR the Midland also controlled a branch line to Ystalyfera and Brynamman, serving several collieries along its length, but this was the only incursion by the Midland into the South Wales coalfield.

Neath & Brecon Railway

Neath to Brecon (opened 1867)

The line was used, in addition to coal transport, as part of a through route for Midland Railway trains between Brecon and Colbren Junction, where the MR-owned Swansea Vale Railway was joined (see above). The Neath & Brecon Railway was absorbed by the GWR in 1922.

Port Talbot Railway

Port Talbot to Pontyrhyll (opened 1898)

Serving more than fifty collieries, the Port Talbot Railway & Docks enlarged the docks at Port Talbot and built a meandering line from here to Maesteg and Pontyrhyll in the Garw Valley, where it met the GWR's line to Blaengarw. Built to transport coal, the PTR&D also had running powers for passenger trains up the Garw Valley to Blaengarw. The PTR&D opened an extension to the Ogmore Valley along the route of two earlier tramways and another extension connected with the South Wales Mineral Railway (see below) at Tonmawr. Although worked by the GWR from 1908 the PTR&D remained independent until 1923.

Rhondda & Swansea Bay Railway

Swansea to Treherbert (opened 1895)

The steeply graded line followed the Afan Valley to Cymmer Afan before reaching Treherbert in the Rhondda Valley via the 3,443-yd-long Rhondda Tunnel. Coal was carried from the Rhondda to the enlarged docks at Swansea, bypassing the Taff Vale Railway (see below) and the Cardiff Docks bottleneck. The line was worked by the GWR between 1906 and 1922 and was closed between Neath and Cymmer Afan in 1962 although the Rhondda Tunnel, the longest in Wales, continued to be used by trains from Bridgend until 1970.

Rhymney Railway

Rhymney Ironworks to Cardiff Bute Docks (opened 1858)

Part of this route was over the Taff Vale Railway (see below) but working arrangement issues led the Rhymney Railway (RR) to build its own route from Caerphilly to Cardiff. Opening in 1871 the new route also granted running powers to the LNWR, which met the RR at the head of the Rhymney Valley. The GWR absorbed the RR in 1922. The entire route between Rhymney and Cardiff via Caerphilly is still open for business.

South Wales Mineral Railway

Briton Ferry to Collieries (opened 1861) and to Glyncorrwg (opened 1863)

Originally built to Brunel's broad gauge, the heavily engineered South Wales Mineral Railway (SWMR) featured a 1-in-10 rope-worked incline, gradients as steep as 1-in-22 and the 1,109-yd-long Gyfylchi Tunnel on its route between Briton Ferry and Glyncorrwg. Converted to standard gauge in 1872, the SWMR was taken over by the GWR-worked Port Talbot Railway & Docks (see above) in 1908, closing in three stages between 1910 and 1970.

Taff Vale Railway

Cardiff Docks to Merthyr Tydfil (opened 1841)

The 24½-mile route was doubled between 1845 and 1861. The TVR went on to lease two other local railways and in 1889 purchased six more. With the purchase of the Cowbridge & Aberthaw Railway in 1894 and the Aberdare Railway in 1902, the TVR ended up with a route mileage of 124½ serving hundreds of collieries in the Rhondda, Taff and Aberdare Valleys. The company became a constituent of the enlarged GWR in 1923. During its lifetime the TVR became one of the most profitable railway companies in South Wales, rewarding its shareholders each year with handsome dividends. The TVR's main lines from Cardiff to Pontypridd thence to Treherbert, Aberdare and Merthyr are still open for business.

LEFT: The LMS incursion deep into the South Wales Valleys – former LNWR 0-6-2T No. 7690 waits to depart from Brynmawr with the 4.30 p.m. Tredegar to Abergavenny train on 7 July 1938.

BELOW: A former GWR tank locomotive hauls an auto train up the Rhondda Valley in 1954. The photograph is a still from the British Transport Films 'Every Valley', a day-in-the-life study of the transport services in this South Wales community.

Excursion trains

During the 1840s the idea dawned upon a few enterprising men, of whom Thomas Cook and the Midland Railway's James Allport became the most famous, that there might be a 'business of travel' in addition to the travel required by business – from this idea grew the excursion train. The first recorded excursion took place in 1831 when the Liverpool & Manchester Railway laid on a special train at reduced fares for members of a local Sunday school. Excursion trains had not only grown popular but had grown in size by the 1840s. On 24 August 1840, the Midland Counties Railway ran an early excursion of gigantic proportion to an exhibition at Leicester from Nottingham. Apparently the enormous train of nearly seventy carriages conveying 2,400 people passed majestically before astonished spectators. There seemed no limit to the size of excursion trains but surely the record must go to one organized by the York & North Midland Railway in 1844 when around 6,500 day trippers were conveyed between Leeds and Hull in over 200 carriages hauled by nine locomotives!

The introduction of Bank Holidays in 1871 opened up more opportunities for railway companies and travel agents to run special trains in which hundreds of thousands of workers and their families were carried from industrial cities to nearby seaside resorts such as Southend, Blackpool, Skegness and Brighton – on Bank Holidays during the early twentieth century, Blackpool's three railway stations were filled to bursting with excursion trains that brought thousands of day trippers from the cotton mill towns of Lancashire.

Horse racing meetings were another big money spinner for the railway companies. Not only did the railways convey the horses in special trains but extra trains were laid on for racegoers during race days. Racecourses at Newbury, Cheltenham, Stratford-upon-Avon, Aintree, Haydock Park, Wetherby, and Hedon even had their own stations. The station at Epsom Downs, home of the Derby, possessed no fewer than nine platforms and numerous sidings all designed to cope with the massive influx of trains on race days.

LEFT: A LNWR poster advertising excursions from northwest England to the 1922 FA Cup Final. Football specials were a regular fixture at weekends on Britain's railways until the late twentieth century.

BELOW: Every dog has his day – a charming poster promoting holiday excursions produced by the Great Northern Railway in 1913, reminding day trippers not to forget the family pet.

In the late nineteenth and early twentieth century monster railway excursions, such as the Messrs. Bass annual trip from Burton-on-Trent and the yearly excursion of GWR employees and their families from Swindon, were run. The brewing firm of Bass often chartered up to seventeen specials and by the time the last of these had left Burton the first had arrived at its destination, Blackpool, ninety miles away! The GWR annual trip from Swindon consisted of a succession of trains run to different destinations departing at 10-minute intervals – in 1904 23,000 people took part (including wives and nearly 10,000 children), conveyed in twenty-one special trains to holiday destinations such as Weymouth, Weston-super-Mare, London, Winchester, Manchester, Birkenhead, South Wales and Devon resorts. Some of the passengers returned the same day while others stayed on holiday for a week and all travelled free. During the annual summer one-week shutdown of the railway works, Swindon was a ghost town.

In the twentieth century one of the most important organizers of excursion trains was the CTAC (Creative Tourists Agents Conference), a consortium of nine UK travel agents that chartered special trains for workers during their annual summer holiday shutdown in the 1930s and from 1945 to 1968. Excursion trains for racegoers, football supporters and holidaymakers also continued to play an important role on Britain's railways until the 1960s but increasing car ownership and changing holiday habits eventually brought an end to 'days out by train'.

Thomas Cook

Thomas Cook was born in Derbyshire in 1808. He was a strict Baptist and a member of the Temperance Society. Following a move to Market Harborough in Leicestershire, Cook organized his first excursion in 1841 when he hired a train from the Midland Counties Railway to take 540 abstainers from Leicester to a temperance rally in Loughborough. Cook went on to arrange more railway outings from Leicester for temperance societies and Sunday schools during the following years, and in 1851 arranged excursions to the Great Exhibition at Crystal Palace for over 150,000 people. His success led to his first foreign adventure in 1855 when he arranged a trip from Leicester to the Paris Exhibition, taking a percentage of the cost of railway tickets. By the 1860s he was arranging tours further afield to Italy, Egypt and the USA – he never looked back. Railway excursions and day trips, especially to the seaside, became a popular part of British social life.

RIGHT: Poster produced by the South Eastern & Chatham Railway in 1910 to promote rail links with round-the-world travel tickets offered by the tour operator Thomas Cook & Son. Thomas Cook established his travel firm in 1841 and was soon offering holidays to destinations around the world.

London's mainline stations

The opening of the Great Railway's London Marylebone terminus in 1899 was the culmination of over six decades of major railway building in London. It would be 2007, when the High Speed 1 Channel Tunnel line opened to St Pancras, before a new railway was to reach London.

One major London station closed in the nineteenth century. Opening in 1844, Bricklayers' Arms was used for just one year by the London & Croydon Railway and by the South Eastern Railway until closure in 1852. It was designed by Lewis Cubitt with a frontage similar to his later work at King's Cross. Despite official closure, Bricklayers' Arms was used on an infrequent basis for excursion traffic until the Second World War.

London Bridge

London Bridge station, London's first terminus, has a complicated history. It consisted of two separate stations, the earliest of which was opened by the London & Greenwich Railway (L&GR) in 1836. An adjacent station was opened by the London & Croydon Railway (L&CR) in 1839 and was also served by the London & Brighton Railway (L&BR) from 1841 and the South Eastern Railway (SER) from 1842. The first station was rebuilt in 1844 for use by the L&CR, the SER and the L&BR with the second station then being used by the L&GR.

The L&CR and L&BR amalgamated in 1846 forming the London, Brighton & South Coast Railway (LB&SCR) – a new enlarged station replaced the former joint station in 1854 and was further enlarged in 1862 when a single-span arched glass roof was added. The original adjacent station was rebuilt and enlarged by the SER in 1850, before closing Bricklayers' Arms terminus. It was further rebuilt in 1864 allowing through trains to run to the new Charing Cross station, and in 1866 to provide access to the new Cannon Street station. London Bridge suffered major damage during the Second World War Blitz and is currently being extensively rebuilt.

Euston

Opening in 1837, the London & Birmingham Railway's Euston terminus was designed by Philip Hardwick, and enlarged by the London & North Western Railway in 1849 with the Great Hall designed by Hardwick's son, Philip Charles. British Railways controversially demolished this historic building and its famous Doric Arch in the early 1960s, replacing it in 1968 with the current featureless building and eighteen platforms serving the West Coast Main Line.

Paddington

The London terminus of Isambard Kingdom Brunel's broad-gauge GWR to Bristol which was completed in 1841. The present overall glass-roofed structure was designed by Brunel and Matthew Wyatt, and opened in 1854. The GWR considerably enlarged the station before the First World War and it suffered bomb damage during the Second World War. Currently only the Heathrow Express service from Paddington is electrified, with full-scale electrification of routes to the west due to be completed in 2017.

Fenchurch Street

One of the smallest of London's main line rail termini, with only four platforms, Fenchurch Street was opened in 1841 by the London & Blackwall Railway, replacing the temporary Minories station opened a year earlier. The station was designed by William Tite but was rebuilt in 1854 with a vaulted roof and new façade designed by George Berkeley. In 1858 it became the London terminus of the London, Tilbury & Southend Railway. Today it serves the electrified lines to Southend.

Waterloo

Designed by William Tite, the original station at Waterloo was built on marshy ground close to the River Thames' south bank by the London & South Western Railway, opening in 1848. It was extended in the 1880s and rebuilt in the early twentieth century before being damaged in the Second World War Blitz. Five platforms were assigned to Eurostar services between 1994 and 2007. With its twenty-two platforms Waterloo is Britain's busiest station.

King's Cross

Designed by Lewis Cubitt and built on the site of a fever and smallpox hospital, King's Cross was opened by the Great Northern Railway in 1852. It replaced a temporary terminus opened at Maiden Lane two years earlier. Recently redeveloped, it is the southern terminus of the East Coast Main Line with twelve platforms, one of which is numbered '0'.

ABOVE: A busy scene on Platform 1 at Paddington station on 10 June 1931. In the right foreground is a GWR 0-6-0 pannier tank that has brought the empty stock of the West of England express train from Old Oak Common sidings. Over eighty years later the scene, apart the changes in fashion and the rolling stock and locomotives, is very much the same.

Victoria

Victoria comprises two stations side by side – the West side was opened by the London, Brighton & South Coast Railway in 1860 and was rebuilt in the early twentieth century. The East side was opened by the London, Chatham & Dover Railway in 1862. For four years it was also served by GWR broad-gauge trains via the West London Extension Joint Railway. The South Eastern & Chatham Railway rebuilt it in the early twentieth century. The combined nineteen platforms came under single ownership in 1923 upon formation of the Southern Railway.

Charing Cross

Designed by Sir John Hawkshaw, the original Charing Cross station was built on the site of Hungerford Market by the South Eastern Railway, opening in 1864. Its overall roof collapsed in 1905 killing six people. The station was redeveloped in the 1990s with a modern office and shopping complex above its six platforms.

Broad Street

Serving the City of London, Broad Street station was opened by the North London Railway in 1865 and extended in 1891 and 1913. It suffered damage during a Zeppelin bombing raid in 1915 and extensive damage during the Second World War Blitz, after which it was not fully repaired. It closed in 1986, one of only two of London's termini to close in the twentieth century. It has since been demolished.

Cannon Street

Designed by John Wolfe Barry and Sir John Hawkshaw, Cannon Street was opened by the South Eastern Railway in 1866. Featuring a single-span glass-and-iron overall roof set between two distinctive towers, the station was joined by an Italianate-style hotel designed by E.M. Barry a year later. The original glass roof was removed before the Second World War and the station suffered severe damage during the Blitz. Its twin towers survived the 1960s redevelopment, being listed as Grade II structures. It was further redeveloped in the early twenty-first century.

St Pancras

The grand London terminus of the Midland Railway, St Pancras, was designed by William Henry Barlow. When it opened in 1868, its single-span overall arched iron-and-glass roof was the largest structure of its kind in the world. The Gothic-style hotel frontage of 1873 was designed by George Gilbert Scott. The station was damaged during the Second World War, narrowly escaping closure in the 1960s thanks to a campaign led by Poet Laureate John Betjeman. Since 2007 it has also been the London terminus of Eurostar services to and from the Continent via the Channel Tunnel. The newly refurbished Midland Grand Hotel was reopened in 2011.

Liverpool Street

Replacing an earlier terminus at Bishopsgate, Liverpool Street was designed by Edward Wilson and was opened by the Great Eastern Railway in 1874. The adjoining Great Eastern Hotel opened in 1884 and the station was considerably extended in 1895. In 1917 the station was the target of an audacious German air raid killing 162 people and injuring 432. Further damage was sustained in the Second World War and during an IRA terrorist attack in 1993.

Holborn Viaduct

Holborn Viaduct was opened as a through station by the London, Chatham & Dover Railway in 1874. Local services connected with the Metropolitan Railway at Farringdon via Snow Hill Tunnel beneath Smithfield Meat Market. In 1916 passenger services ceased, effectively making Holborn Viaduct a commuter terminus. Freight traffic continued through the tunnel until the late 1960s. The tunnel reopened for Thameslink services in 1988 but little-used Holborn Viaduct station closed in 1990.

Blackfriars

In 1886 the London, Chatham & Dover Railway (LC&DR) opened St Paul's, a through station connecting with the company's stations at Ludgate Hill, Holborn Viaduct and the Metropolitan Railway at Farringdon via Snow Hill Tunnel. Mainly a commuter station, it was renamed 'Blackfriars' in 1937 and rebuilt in the 1970s. Blackfriars has recently been upgraded and half of its energy requirement is now supplied by 4,400 solar panels fitted to Blackfriars Railway Bridge.

Marylebone

Designed by Henry Braddock, Marylebone station was opened as the southern terminus of the Great Central Railway in 1899. The GCR's through route was closed in 1966 but Marylebone remained open, serving Aylesbury and the Chilterns. It is the only London terminus that is not electrified. Despite threats of closure in 1983 it has survived and serves as the terminus of Chiltern Railway services to and from Aylesbury, Birmingham and Oxford.

LEFT: Early morning rush hour at Liverpool Street station, 12 October 1951, at that time the busiest terminus in London. In 1949 the route to Shenfield was electrified to provide a more efficient and faster service, although the station still remained very crowded at peak times.

Famous locomotive engineers

William Dean

The son of a soap factory manager, William Dean was born in London in 1840 and following his education became an apprentice engineer under Joseph Armstrong at the Great Western Railway's Stafford Road Works in Wolverhampton. In 1863 Dean was appointed as Armstrong's chief assistant – a year later Armstrong was appointed as the GWR's Chief Locomotive Engineer at Swindon Works following the retirement of Daniel Gooch. At Wolverhampton George Armstrong replaced his brother Joseph and Dean was promoted to manager of the works. Dean's rise up the corporate ladder continued in 1868 when he was appointed as Joseph Armstrong's chief assistant at Swindon.

Joseph Armstrong's sudden death in 1877 saw Dean catapulted into the top post at Swindon and over the next twenty-five years he was responsible for designing many successful locomotive classes for the GWR including the '3300' Class 4-4-0 'Bulldog', the '3252' Class 4-4-0 'Duke', the '2600' Class 2-6-0 'Aberdare' and his long-lived '2301' Class 0-6-0 'Dean goods'. During his time at Swindon Dean had successfully overseen the smooth transition from broad gauge to standard gauge, a process that was completed in 1892. Following several years of illness Dean retired in 1902, to be succeeded by G.J. Churchward, but died three years later in Folkestone.

ABOVE: William Dean's elegant 4-2-2 locomotive No. 3050 *Royal Sovereign* is seen here at speed on 23 September 1898 soon after leaving Box Tunnel with the 7.15 a.m. express from Falmouth to London Paddington. With their 7-ft 8-in. driving wheels and polished brass and copper fittings, the 'Dean Singles' epitomized GWR express steam at the end of the century.

Henry Ivatt

Born in Cambridgeshire in 1851 Henry Alfred Ivatt joined the London & North Western Railway (LNWR) at the age of 17 as an apprentice to John Ramsbottom, the locomotive superintendent at Crewe Works. After working as a fireman Ivatt was then put in charge of the locomotive depot at Holyhead in 1874 before being promoted to the LNWR's Chester District. Three years later Ivatt moved to Ireland to work for the Great Southern & Western Railway at Inchicore Works in Dublin, where in 1882 he became chief locomotive engineer. His claim to fame in Ireland was to patent a sprung flap for vertically opening carriage windows, which was widely used on Britain's railways until the 1960s.

In 1895 Ivatt returned to England and was appointed Locomotive Superintendent of the Great Northern Railway at Doncaster Works. Here he designed many successful locomotive classes including the Class 'C1' 'Klondyke' 4-4-2s of 1898, the first application of this wheel arrangement on a tender engine in Britain. A total of twenty-two of these express locomotives were built and were followed by eighty of his larger boiler versions, some of which remained in service until the 1950s. Succeeded by Nigel Gresley, Henry Ivatt retired in 1911 and died in 1923.

ABOVE: One of Henry Ivatt's Class 'C1' 4-4-2 'Atlantics' is seeing thundering through Potters Bar with the 'Leeds Pullman' in 1932. Seventeen of these fine locomotives survived into the BR era with the last withdrawn in 1950. One has been preserved and is on display at the National Railway Museum in York.

Samuel Johnson

Born in Yorkshire in 1831, Samuel Waite Johnson was educated at Leeds Grammar School before following an engineering career with the Leeds' locomotive builders of E. B. Wilson & Co. At the age of 28 he was appointed Acting Locomotive Superintendent of the Manchester, Sheffield & Lincolnshire Railway then, in 1864, as Locomotive Superintendent of the Edinburgh & Glasgow Railway. A further move came only two years later when Johnson was appointed Chief Mechanical Engineer (CME) of the Great Eastern Railway at its Stratford Works in East London. Here he designed twelve classes of locomotive including the numerous Class '417' and Class '477' 0-6-0s before finally moving to Derby where he became CME of the Midland Railway. His time at Derby was marked by the design of many highly successful classes including over 900 of the 0-6-0 freight locos, some of which remained in service until the 1960s. His elegant 4-2-2 '115' Class 'Spinners' were capable of hauling express trains at speeds of over 80 mph while his 3-cylinder compound '1000' Class 4-4-0s, introduced in 1902, were the mainstay of Midland Railway's express passenger trains well into the 1930s, some remaining in service until the 1950s. Succeeded by Richard Deeley, Johnson retired in 1903 and died in 1912.

ABOVE: Nicknamed 'Spinners' for their turn of speed and designed by Samuel Johnson for the Midland Railway, 4-2-2 locomotive No. 673 is now on display at the National Railway Museum in York.

William Stroudley

Born in Oxfordshire in 1833, William Stroudley was apprenticed at the age of 14 to the engineering company of John Inshaw in Birmingham where he gained experience on steam engines. In 1854 Stroudley moved to the GWR's Swindon Works as a trainee locomotive engineer under Daniel Gooch before working for the Great Northern Railway at Peterborough. A move to Scotland came in 1861 when he was appointed Works Manager at the Edinburgh & Glasgow Railway's Cowlairs Works in Glasgow. Four years later he moved to Inverness where he was appointed as Locomotive and Carriage Superintendent at the Highland Railway's Lochgorm Works – due to financial constraints he was only able to design one locomotive class here, the diminutive 0-6-0ST being the forerunner of his later 'Terrier' locos for the London, Brighton & South Coast Railway (LB&SCR).

Stroudley's final move came in 1870 when he was appointed Locomotive Superintendent of the LB&SCR at its Brighton Works following the resignation of John Craven. During his tenure he designed many successful and graceful locomotive classes ranging from the 'A1' Class 0-6-0Ts, or 'Terriers' as they became known', of which fifty were built with some remaining in service until 1964, the 125 'D1' Class 0-4-2T suburban passenger locos and the 80 'E1' Class 0-6-0T goods locos. Without doubt the highpoint of his career, the 36 'B1' Class 0-4-2 express passenger locomotives were the mainstay of the London to Brighton services – they were all named after politicians and were nicknamed 'Gladstones' after the first to enter service in 1882. Sadly, Stroudley died from bronchitis while exhibiting one of these fine locomotives at the Paris Exhibition in 1889. He was succeeded at Brighton by R.J. Billinton.

ABOVE: Designed by William Stroudley for the London, Brighton & South Coast Railway and introduced in 1872, the 'Terrier' Class 'A1' 0-6-0Ts had a long working life. Here No. 32678 is seen hauling a train on the Hayling Island branch in the late 1950s.

1901	1902	1903	1904	1905	1906	1907	1908	1909	1910	1911
Mallaig Extension opens		GWR operates first motor bus route between Helston and The Lizard	GWR's *City of Truro* allegedly achieves 100 mph	Lots Road Power Station opens in London	28 killed in train derailment at Salisbury	Wath Marshalling Yard opened by Great Central Railway		LB&SCR introduces suburban electric trains from Victoria station	LSWR establishes locomotive works at Eastleigh	Rhyl Miniature Railway opens

THE EARLY TWENTIETH CENTURY

1901–1922

1912	1913	1914	1915	1916	1917	1918	1919	1920	1921	1922
	LSWR transports 120,000 troops of British Expeditionary Force to Southampton	Wartime Railway Executive Committee takes control of Britain's railways	227 people killed in multiple train collision at Quintinshill		War Department Light Railways builds 700-mile network of lines in Northern France			Government control of railways ends	15 people killed in train collision at Abermule in mid-Wales	

The monopoly of the North Eastern Railway

Recovering from the downfall of the disgraced 'Railway King', George Hudson, the North Eastern Railway (NER) was formed in 1854 by four constituent companies: York, Newcastle & Berwick Railway, York & North Midland Railway, Leeds Northern Railway and the Malton & Driffield Railway. At its formation the NER was the largest railway company in Britain owning 720 route miles along with various canals and dock installations. The company went on to totally dominate northeast England by absorbing twenty-five more railways and three dock companies between 1857 and 1922, giving it a network of just under 5,000 route miles before it became one of the major constituent companies of the London & North Eastern Railway in 1923.

The as-yet-unbuilt route northwards from Darlington to Gateshead via the Team Valley was reinstated by the NER in 1865 and opened throughout in 1872, and the ECML between York and Newcastle that we know today was complete. At its northern end, Newcastle Central station, reached over the Tyne via Robert Stephenson's High Level Bridge, had already been opened in 1850 by the York, Newcastle & Berwick Railway and the Newcastle & Carlisle Railway. However, the small terminus station at York was woefully inadequate, with trains from London to Newcastle required to reverse out of the station before continuing their journey northwards. With its thirteen platforms and overall curving arched roof, the present station at York was opened by the NER in 1877 and, at the time, was the largest station in the world.

From its inception in 1854 the NER had set up its headquarters at Gateshead in Newcastle, where the company established its own locomotive, carriage and wagon works. Under successive locomotive superintendents – chiefly Edward Fletcher and Wilson Wordsell – the works turned out many classic steam locomotive types and in 1904 introduced the ground-breaking electrified North Tyneside suburban railway. Wordsell was replaced by Vincent Raven (see box feature) in 1910 by which time locomotive construction had been transferred to Darlington. Gateshead Works closed completely in 1932.

Until 1923 the company had a total monopoly on the railway routes serving the collieries, iron ore mines and iron- and steel-making regions of Northumberland, County Durham and Yorkshire – in its heyday before the First World War, the NER carried a larger tonnage of coal and mineral traffic than any other railway in Britain. Its major dock installations at Tyne Dock, Hartlepool, Middlesbrough and Hull handled vast amounts of goods ranging from coal, iron and steel exports and grain, wood and fruit imports. The company also operated its own fleet of steamships for the North Sea trade with Scandinavia and northern Europe.

PREVIOUS SPREAD: Women played a vital role on Britain's railways during both world wars. Apart from driving and firing steam locomotives they were hard at work all over the country's railway network in jobs that were previously considered the preserve of men. Here a group of women locomotive cleaners pose for the camera, sitting on Lancashire & Yorkshire Railway 'High Flyer' Class 4-4-2 locomotive No. 1406 at Low Moor engine shed near Bradford, 23 March 1917.

BELOW: This 1907 poster was produced by the North Eastern Railway to promote its service between Hull and Edinburgh, journey time an incredibly slow five hours! Sadly this particular journey is no longer possible as the southern part of the route between Beverley and York was closed in 1965.

Vincent Raven

Several impressive designs of steam locomotive marked Vincent Raven's tenure as Chief Mechanical Engineer for the NER including his Class 'Z' 4-4-2s and five Pacific locos of his design (three being built under LNER management in 1924) which were all named after northeastern cities. Featuring prominently were heavy freight locomotives as the NER carried more coal (and vast tonnages of iron ore and steel) than any other railway in Britain. His development of Wordsell's 0-8-0s (LNER Class 'Q6' and 'Q7') stood the test of time, surviving well into the BR era. What is not so well known is that Raven advocated the electrification of the York to Newcastle mainline, having previously electrified in 1925 the Shildon to Newport coal-carrying line. A prototype mainline electric loco had been built at Darlington in 1922 years ahead of its time, but Raven's plan was dropped the following year by the LNER.

ABOVE: Designed by Vincent Raven for the North Eastern Railway and introduced in 1913, the 'Q6' Class 0-8-0 heavy freight locomotives had a long working life. Here No. 63346 drifts down towards Seaton with a train from South Hetton Colliery to Dunston Staithes on 22 October 1966.

Early overground electrification schemes

The early twentieth century saw the introduction of six pioneering electrification schemes on Britain's overground railways. With their faster acceleration and cleanliness, the new electric trains were a great success and a forerunner of today's modern commuter trains.

North Eastern Railway

By the end of the nineteenth century it had become rather fashionable for Newcastle businessmen to commute to work from their elegant homes on the coast. Resorts such as Whitley Bay and Tynemouth were within easy reach of the city centre along the existing steam-operated network but by 1903 competition from new electric street trams had seen railway passenger numbers in steep decline. Faced with this loss of income the North Eastern Railway (NER) took the bold decision to fight back by electrifying a near-circular route out from Newcastle Central station via the coastal towns of Tynemouth, Whitley Bay and Monkseaton to the former Blythe & Tyne Railway's terminus at Bridge Street. The railway was electrified with a third rail at 600V DC and new electric multiple units, built at the NER's workshops in York, began operating services on 29 March 1904.

Lancashire & Yorkshire Railway

Beating the North Eastern Railway by just one week the Lancashire & Yorkshire Railway (L&YR) became the first in Britain to operate a suburban electric service. A fourth rail system was used at 600V DC with the first section opening between Liverpool Exchange and Crossens via Southport on 22 March 1904. With routes subsequently opened to Aintree, Meols Cop and Ormskirk the network of the L&YR's suburban electric lines radiating out from Liverpool had reached thirty-seven miles by 1913.

The L&YR also electrified the Bury to Holcombe Brook branch line in 1913 but this time an overhead 3.5kV DC system was used until conversion to third rail in 1918. Electrification of the company's Manchester to Bury route was also completed in 1916.

Midland Railway

While the Midland Railway (MR) experimented successfully with electrification of its route from Lancaster Green Ayre to Morecambe and Heysham in 1908, this line remained unique in the company's network. Like the London, Brighton & South Coast Railway's electric commuter lines that opened a year later in South London, the MR used an overhead power supply at 25-cycle 6,600V AC. The original electric multiple units remained in service until 1951 after which the route was converted to 50-cycle 6,600V AC. Later used as a test bed for the standard 25kV 50-cycle system, the Lancaster Green Ayre to Morecambe line closed in 1966.

London & North Western Railway

With links to the District Railway and the new extension of the Bakerloo Line, the London & North Western Railway (LNWR) embarked on a major electrification scheme for its inner-suburban network in North London that included lines from Broad Street to Richmond and Euston to Watford. With electricity generated at the company's Stonebridge Park power station the 630V DC fourth rail system opened in stages between 1914 and 1922.

London, Brighton & South Coast Railway

The early years of the twentieth century found the London, Brighton & South Coast Railway (LB&SCR) suffering from an acute shortage of steam locomotives caused by an inefficient locomotive department at Brighton Works. In the light of this the electrification of its intensely operated suburban services in South London had been mooted in 1900 but it would be 1909 before the first route, between Victoria and London Bridge stations, was up and running. The overhead power supply for London's first overground electric railway was the same as that used on the Midland Railway's new route from Lancaster to Morecambe, 25-cycle 6,600V AC. The new route, dubbed the 'Elevated Electric' was a great success and other lines soon followed but the outbreak of the First World War delayed the electrification of the rest of the LB&SCR's South London suburban lines. This was only completed in 1921 by which time there were 24½ route miles in place. Despite plans to extend electrification on the mainlines to the south coast the London & South Western Railway's third-rail system eventually won over under new Southern Railway management and the last 'Elevated Electric' train ran in 1929.

London & South Western Railway

Faced with a decline in passenger numbers on its southwest London suburban network in the early twentieth century the London & South Western Railway (LSWR) lagged behind before it, too, implemented its own electrification scheme. Unlike the LB&SCR (see above) the LSWR opted for a third-rail system with a current of 600V DC, the same as the North Eastern Railway's electrified lines in Tyneside. With great success electric multiple units started operating on the inner suburban services out of Waterloo in 1915 and by the outbreak of the First World War had been extended to Claygate in Surrey, making a total of fifty-seven route miles. Following the 'Big Four' Grouping of 1923 the new Southern Railway chose the LSWR's third-rail system for its massive mainline electrification programme.

LEFT: Designed by Vincent Raven for the North Eastern Railway, 1,500V DC Bo-Bo electric locomotive No. 8 is seen on trial with a dynamometer car at Newport, 12 October 1921. Introduced in 1914, this loco was one of ten used to haul coal trains on the 18-mile route between Shildon and Newport.

BELOW: The Midland Railway electrified its route between Lancaster, Morecambe and Heysham in 1908. Here, brand new electric coach No. 2236 is seen at Heysham shortly after the line was 'switched on'.

Sleeping cars

The London & North Western Railway introduced the first sleeping car in 1873 on the West Coast Main Line between London Euston and Glasgow Central. Other companies were soon following suit using imported Pullman cars with convertible seating – the Midland Railway led the way by buying the luxury carriages in kit form from George Mortimer Pullman's company in the USA. However, all sleeping arrangements on trains were communal until the introduction of Great Western Railway's purpose-built sleeping car. Introduced in 1890 and equipped with double-berth cabins these were similar to those still used today.

Sleeping car trains were running from London to Scotland, the West Country, North Wales and Northern England by the turn of the century. Apart from the wartime periods these services remained more or less intact until the 1960s, by which time British Railways was on most nights running around forty trains.

Following the 'Big Four' Grouping of 1923, the London Midland & Scottish Railway became the largest operator of sleeping car trains from London to numerous destinations: Barrow-in-Furness, Carlisle, Edinburgh Waverley, Galashiels, Glasgow Central, Glasgow St Enoch, Holyhead, Inverness, Leeds City, Liverpool Lime Street, Manchester Piccadilly, Motherwell, Oban, Perth, Preston, Stranraer Harbour and Whitehaven Corkickle. Additionally there were services between Birmingham and Glasgow, Newcastle and Bristol, Edinburgh Princes Street and Birmingham, Glasgow and Liverpool, Glasgow and Manchester, and Manchester and Plymouth.

Sleeper trains were operated by the London & North Eastern Railway along the East Coast Main Line from London: to Aberdeen, Arbroath, Dundee, Edinburgh Waverley, Fort William, Montrose and Newcastle.

There were no sleeping car trains on the Southern Railway apart from the cross-Channel services to Paris and Brussels on 'The Night Ferry' operated by Compagnie Internationale des Wagons-Lits. Services from Paddington to Birkenhead Woodside, Swansea and Carmarthen, Penzance, Plymouth and between Plymouth and Manchester were operated by the Great Western Railway.

Britain's network of sleeping car trains has shrunk considerably since the heady days of the 1930s with the advent of cheaper domestic air travel and high-speed motorways. Today only two routes remain in operation, the 'Caledonian Sleeper' and the 'Night Riviera', both of which operate six nights each week and are among the last locomotive-hauled passenger trains in Britain. The former operates out of London Euston and is run in two portions, the first with carriages for Aberdeen, Fort William and Inverness and the second with carriages for Edinburgh and Glasgow. The 'Night Riviera' runs between London Paddington and Penzance.

BELOW: Poster produced in 1928 for the London Midland & Scottish Railway's third-class sleeping car service between London and Scotland. There are sadly now only two sleeping car routes left operating in Britain, one from London to various destinations in Scotland and one from London to Penzance.

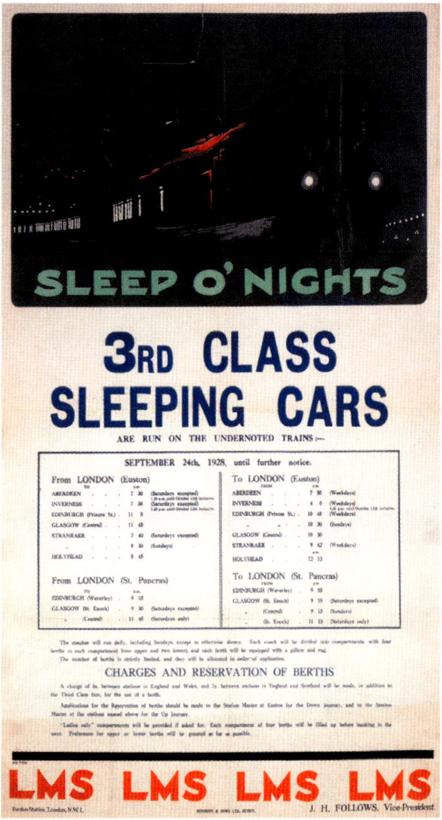

Restaurant cars

Dining cars were first introduced on Britain's railways in 1879 when the Great Northern Railway introduced the first vehicle of its kind in the country on its London King's Cross to Leeds service. Soon following suit in 1882 was the Midland Railway, which put dining cars on its competing London to Leeds service and within a short time the three competing routes from London to Manchester operated by LNWR, MR, and GNR all had similar accommodation. Dining cars on the longer Anglo-Scottish routes of the East and West Coast mainlines only started to appear in 1893 due to the vested interests of refreshment rooms at Preston, York and Normanton stations where passengers usually alighted to take lunch before resuming their journey.

A 99-year agreement had been made with a catering company in 1842, putting the GWR in a difficult situation as all their trains between Paddington, Bristol and South Wales were forced to stop at Swindon for ten minutes for refreshments. This intolerable situation resulted in the GWR buying out the Swindon Junction Hotel Company for £100,000 in 1895, after which dining cars soon appeared on GWR routes.

By the early twentieth century the provision of dining cars was greatly facilitated by the introduction of corridor trains – this allowed more passengers to be fed in comfort with less plant and staff. On short journeys it was often necessary to provide accommodation for feeding a large number of passengers simultaneously – the Great Eastern Railway's 'Harwich Boat Express' occupied a journey of only an hour and a half but was able in that time to serve meals to 111 passengers from a kitchen in the centre of the train measuring only 6 ft by 17 ft. While the greater part of the food consumed in restaurant cars was cooked in the tiny confines of the on-train kitchen, some items such as soup, pastry and sweets were usually prepared at railway-owned hotels beforehand. At Liverpool Street the Great Eastern Hotel baked bread and pastry for restaurant cars and refreshment rooms down the GER main line, while the LNWR had a bakery at Euston and Rugby and the GNR at Peterborough. The railway restaurant car business was booming by 1904 – the LNWR/Caledonian Railway served in its West Coast Joint vehicles between Euston and Glasgow over 500,000 meals annually.

Nearly 140 years after being introduced in Britain the last few surviving restaurant car services are to be found on the East Coast Main Line between King's Cross and Edinburgh and on the First Great Western Pullman services between Paddington and Plymouth and Paddington and Swansea. Passengers on other trains are offered a meagre selection of expensive snacks and drinks that are wheeled up and down the train on airline-style trolleys.

BELOW: How times have changed! A London & North Eastern Railway poster produced in the 1930s promotes the 200 restaurant cars operating on the company's trains. The 'then and now' images compare modern rail catering with that of 100 years before.

Railway towns

The railways were one of Britain's main employers by the early twentieth century. Of the estimated 600,000 employees, about one third were engaged in the maintenance and renewal of the permanent way and rolling stock. In the latter category tens of thousands were employed at strategically located locomotive, carriage and wagon works around the country. Although many of them, such as the Great Northern Railway's Doncaster Works, the Midland Railway's Derby Works and the Great Eastern Railway's Stratford Works, were located in existing towns and cities, many new establishments were also built on greenfield sites. One-time small villages and former railway towns such as Ashford, Crewe, Eastleigh and Swindon owe their existence today to the fact that they were once great railway centres. These paternalistic railway companies looked after their employees from the cradle to the grave, decades before the introduction of the welfare state.

Horwich

By the end of the nineteenth century, Horwich, in Lancashire, was one of the youngest railway towns to come into existence. When in 1887 the Lancashire & Yorkshire Railway's mechanical engineering works was established at Horwich, the population of the small town was under 4,000; by 1904 it was 16,000, of whom around 10,500 were dependent on the employment provided by the railway company. Private enterprise provided housing for this influx but the railway companies invested in well-equipped technical and leisure institutes for their employees. The Horwich Institute was built with a grant of £5,000 from the L&YR, with a gift from a director's widow paying for additional mechanical and engineering laboratories, and a gymnasium. Another director paid for a cottage hospital. In 1890 about eleven acres of land were given by the railway company for use as a recreation ground for associated cricket, football, bowling and tennis clubs.

Eastleigh

In 1891 the London & South Western Railway (LSWR) had opened a carriage and wagon works at Eastleigh and the small village had grown into a railway town of 9,000 inhabitants in the space of fifteen years. At the same time the LSWR had purchased 200 acres of land adjoining the existing works where they opened a new locomotive works in 1910, transferring about 2,000 more men and their families from the old works at Nine Elms in London.

Corkerhill

Corkerhill, near Glasgow, consisted of one farm and a few cottages until the coming of the railway. Houses in

BELOW: By 1895 when this photograph was taken the London & North Western Railway's locomotive works at Crewe had grown enormously since they had opened in 1843. So what was once a greenfield site had by now become a railway town of some 50,000 souls.

railway towns were not normally owned by railway companies but built by private enterprise and rented to the employees. One exception was the Glasgow & South Western Railway, which in 1896 built a model village at Corkerhill for its locomotive staff and families. The population of 800 people was lodged in 120 houses, and a Railway Institute and allotments were later added. Religion was also important and Sunday services were conducted by members of the Railway Mission in the large hall of the Institute. There was also a Sunday School at Corkerhill, with an average attendance of 110 children, and a Bible Class for young men and women, with an average attendance of seventy.

Ashford

The South Eastern Railway's mainline from London to Dover via the small town of Ashford opened throughout in 1844. The expanding railway soon found its original railway works at New Cross too cramped, so the company set about finding a new green-field site in Kent in 1846. Ashford was chosen because of its convenient location on the new main line. Buying 185 acres of land for £21,000, the SER built a new railway works along with 132 labourers cottages around a village green, a pub, a Mechanics Institute, gas works, a school and public baths.

The first new mainline locomotive was rolled out of the 26½-acre Locomotive Works in 1853. The adjoining Carriage & Wagon Works opened in 1855. Between 1841 and 1861 the population of Ashford more than doubled to 7,000 – the railway was employing around 1,300 workers by 1882.

After the 1899 amalgamation of the South Eastern Railway and the London, Chatham & Dover Railway, Ashford became the main works for the newly formed South Eastern & Chatham Railway.

Swindon

The Great Western Railway established its locomotive, carriage and wagon works at Swindon in 1841. Located at the junction with the branch to Cheltenham on the GWR's broad-gauge mainline between Paddington and Bristol, Swindon was then just a small village but within sixty years had grown into a company-built town of 50,000 inhabitants, 13,000 of whom were directly employed by the GWR. The centre of the town's educational and social activities was the GWR Swindon Mechanics Institute, which was founded in 1844 and provided members with libraries, reading rooms, games rooms, a large hall for musical and dramatic entertainment, a lecture hall and classrooms. Once a year the GWR provided free excursions for members to various destinations – in the summer of 1904 over 23,000 members took part with the company laying on twenty-one special trains. The GWR created its Medical Fund Society in 1847, some one hundred years before the NHS was introduced, membership of which was compulsory for all staff employees in Swindon.

Crewe

Established by the Grand Junction Railway in 1843 the London & North Western Railway's (LNWR) locomotive and steel works at Crewe was originally built on a greenfield site at an important railway junction. The population of this railway town had grown to 40,000 by the early twentieth century with 8,000 being directly employed in the works. The Crewe Mechanics Institute provided evening classes for reading, writing, arithmetic and mechanical drawing with teachers who were also employed at the works. The paternalistic LNWR established a small hospital at Crewe in 1863, which provided free medical care for all its employees, and built schools, churches and homes for its key employees. At its peak in the 1930s Crewe Works employed over 20,000 people and by its closure in 1990 had built a total of over 8,000 locomotives.

Locomotive builders to the world

Britain had established an overwhelming lead in the design and manufacture of steam locomotives since the early nineteenth century and by the early twentieth century there were over fifty independent locomotive and rolling stock manufacturers in Britain – these were in addition to the twenty establishments that were owned by the major railway companies. While the latter concentrated on producing locomotives for their own British railway companies the former had also gone on to become major suppliers for railways around the world and in particular the British Empire.

Many of these manufacturers had been established in the heady days of Britain's railway revolution and were concentrated mainly in the Scottish industrial belt, northwest and northeast England, and the Midlands. Companies of note include the following: Andrew Barclay Sons & Co. of Kilmarnock; Avonside Engine Company of Bristol; Birmingham Railway Carriage & Wagon Co.; Dübs & Co. of Glasgow; Gloucester Railway Carriage & Wagon Co.; Kerr, Stuart & Co. of Stoke-on-Trent; Kitson & Co. of Leeds; Manning Wardle of Leeds; Neilson & Co. of Glasgow; Peckett & Sons of Bristol; Robert Stephenson & Co. of Newcastle (founded 1823); Sharp, Stewart & Co. of Glasgow; Vulcan Foundry of Newton-le-Willows; and Yorkshire Engine Co. of Sheffield. However by the early nineteenth century the two largest locomotive manufacturers in Britain were Beyer-Peacock of Manchester and the North British Locomotive Company of Glasgow.

Beyer-Peacock

The locomotive building company Beyer-Peacock was founded in 1853 by English engineer Richard Peacock, German-born Charles Beyer and Scottish engineer Henry Robertson. With its extensive works at Gorton Foundry, Manchester, the company went on to build 4,753 tender locos and 1,735 tank locos for countries ranging from India, Sweden, Spain, Egypt, Turkey, Belgium, Holland, Germany and Italy to Peru, Brazil, Uruguay, Dutch East Indies and the Australian states as well as for many British railway companies. They also built 1,115 Beyer-Garratt articulated locos between 1909 and 1958 for Australia, India, Brazil, Burma, Chile, Ecuador, Brazil, Russia, New Zealand, Sierra Leone, South Africa, Kenya, Uganda, Rhodesia, Nigeria, Iran, Sudan, Angola and last, but not least, the LMS and the LNER in Britain. Of note was the largest steam engine built in Europe – with a tractive effort of 90,000 lb the 4-8-2+2-8-4 Garratt was supplied to Russian Railways in 1932 – and the world's most powerful narrow-gauge locos – the 3-ft 6-in.-gauge Garratts supplied to South African Railways in 1929. During the Second World War the company produced tanks and shells alongside continuing loco production – a Garratt for Burma Railways was designed and built in a record time of 118 days!

By 1958 steam engine production was at an end and Beyer-Peacock went on to build electric and diesel main line locos for British Railways. However this changeover to modern traction proved uneconomic for Beyer-Peacock and the company was forced to close in 1966.

North British Locomotive Company

Formed in 1903 by the merger of three Glasgow locomotive manufacturers – Dübs, Neilson and Sharp Stewart – the North British Locomotive Company became the largest locomotive manufacturer in Europe with a capacity to build 600 per year. With its headquarters in Springburn the NBL went on until the mid-1950s to build thousands of steam locos not only for British companies such as the SR and LMS, but also for the War Department in the Second World War and railways in Australia, New Zealand, Malaysia, Pakistan, India and South Africa – some of these locos have since been preserved and are still in steam.

Sadly the NBL failed to maintain its long and illustrious record when it moved from steam to diesel production in the late 1950s. After signing a deal with the German company MAN to build diesel engines under licence it went on to build some of the most unreliable locomotives ever ordered by BR. The poor workmanship and unreliability of NBL locos brought the company to its knees. Warranty claims and guarantees finally pushed the company over the brink and it went bankrupt in April 1962.

RIGHT: 'Three in four new British locomotives go oversees' – constructing new steam locomotives for the Indian State Railways at the North British Locomotive Company's works in Glasgow, 1946. 75 per cent of the company's locomotives were exported around the world but the changeover from steam to diesel traction brought the company to its knees and it went bankrupt in 1962. Now Britain imports all its railway locomotives from North America and Europe – so much for progress!

Light railways

ABOVE: Lybster station, the southern terminus of the Lybster Light Railway, in 1928. This 13½-mile light railway was opened between Wick and Lybster by the Highland Railway in 1903 but closed in 1944. The locomotive is 4-4-0T No. 15013, one of five built for the Uruguay Eastern Railway between 1891 and 1893 – the order was cancelled so the Highland Railway bought them and nicknamed them 'Yankees'.

Prior to the Light Railways Act of 1896, all new railway lines in Britain required a specific Act of Parliament before they could be built. This was an expensive and time-consuming affair – the severe economic recession of the 1870s and 1880s and associated high interest rates had brought to a halt many new railway schemes in Britain. Railway companies could build new railways in sparsely populated regions of the country without requiring a specific Act of Parliament with the Light Railways Act in place. Although cutting out much of the red tape and expensive legislation it did impose severe limitations including a weight limit of twelve tons per axle and a speed limit of 25 mph. Cost-cutting measures included minimal earthworks, bridges and stations, lightly laid track spiked directly onto sleepers, the absence of level crossing gates, minimal signalling and operating on the 'One Engine in Steam' principal – requiring only one locomotive to work on the line at any given time. Between 1896 and 1925 over thirty standard-gauge and narrow-gauge lines were built under the Act's provisions. Colonel Holman Fred Stephens managed sixteen of these from his modest office in Tonbridge, Kent (see box feature).

While the provisions of the Light Railway Act were perfect for the sparsely populated regions of Scotland, very few light railways were actually built in the country. Of those that saw the light of day were the Dornoch Light Railway and the Leadhills & Wanlockhead Light Railway, both of which opened in 1902, and the Wick & Lybster Railway, which opened in 1903. At the southern tip of the Mull of Kintyre the narrow-gauge Campbeltown & Machrihanish Railway (formerly a canal) was converted from a coal-carrying line to a light railway in 1906. North of Inverness construction of the Cromarty & Dingwall Railway was started but never completed.

By the end of the First World War the building of light railways in Scotland was back on the political agenda and the question of rural transport in Scotland was studied by a Parliamentary Committee which published its findings in 1919. The Committee's report advocated the building of 382 miles of new railway, eighty-five miles of new road, road improvements and the introduction of new bus and steamer services. Schemes looked at were railways to serve the northwest fishing ports of Ullapool and Lochinver, narrow-gauge lines on the islands of Lewis, Skye and Arran, an isolated railway from Dunoon to Strachur and a 28-mile line across the wilds of Galloway linking Dalmellington with the Portpatrick & Wigtownshire Joint Railway at Parton. As all these proposed schemes would pass through sparsely populated and remote countryside the Committee pressed the point that Government funding would be necessary. Not surprisingly, during this period of austerity and increased competition from road transport none of these lines were built.

Today's growing number of heritage railways are also operated as light railways, while on the national network the 'Heart of Wales Line' between Craven Arms and Llanelli has been operated under a Light Railway Order since 1972.

Colonel Holman F. Stephens

The son of Pre-Raphaelite artist Frederic George Stephens, Holman Fred Stephens was born in 1868 and went on to become the leading supporter of the building and management of light railways in England and Wales. As a student he studied civil engineering at University College in London before becoming an apprentice engineer with the Metropolitan Railway. He was appointed assistant engineer for the Cranbrook & Paddock Wood Railway in 1892.

Stephens also had a parallel military career, rising through the ranks to Lieutenant Colonel in the Royal Engineers (Territorial Reserve) by 1916. Seeing a potential business opportunity, he formed the Light Railway Syndicate a year before the Light Railways Act of 1896 had been passed. By the time of his death in 1931 he had control of an empire of light railways across England and Wales. Using second-hand equipment and with the minimum of maintenance these ranged from narrow-gauge enterprises such as the Ffestiniog Railway, the Snailbeach District Railway, the Rye & Camber Tramway and the Ashover Light Railway to meandering standard-gauge rural lines such as the Kent & East Sussex Railway, the Shropshire & Montgomeryshire Railway, the Weston, Clevedon & Portishead Railway and the East Kent Railway. Only three of his standard-gauge lines survived into British Railways ownership in 1948 and these were soon closed. Fortunately for us today one of these, the Kent & East Sussex Railway, is a flourishing heritage railway operating restored vintage steam-hauled trains along the Rother Valley in Kent.

ABOVE: The Kent & East Sussex Light Railway, reopened in 1974, was a typical Colonel Holman F. Stephens branch line, built and operated on a shoestring. Here Manning Wardle 0-6-0 saddle tank *Charwelton* is seen between Wittersham and Rolvenden in 1998 with a period mixed train so typical of the early 1920s.

Railway shipping services

As an island nation Britain has been dependent on seaborne traffic for trade and communications with the rest of the world for hundreds of years. By the nineteenth century, with the Industrial Revolution in full swing, the rapidly expanding growth in manufacturing had become almost totally reliant on railway operations, with imported raw materials and exported finished goods all transported from docksides to factories and vice versa by rail.

The Great Grimsby & Sheffield Junction Railway introduced the first railway passenger shipping service in 1846 when it started operating a ferry across the Humber Estuary between Hull and New Holland – this remained in operation until the opening of the Humber Bridge in 1981. In 1848 the Chester & Holyhead Railway started operating the Holyhead to Kingstown (Dun Laoghaire) packet service to carry the Irish Mail and in 1853 the South Eastern Railway began the first cross-Channel packet service to France between Folkestone and Boulogne. By the early twentieth century railway-operated passenger shipping services had become so successful that there were routes across the North Sea to Scandinavia, Belgium and Holland, across the English Channel to France and Belgium and the Channel Islands and across the Irish Sea to Ireland.

Ireland could be reached via several different routes: Fishguard to Rosslare (GWR); Holyhead to Dun Laoghaire (LNWR); Liverpool to Dublin and Isle of Man (LNWR); and Stranraer to Larne (Glasgow & South Western Railway). The Midland Railway operated the Heysham to Isle of Man route while the Lancashire & Yorkshire Railway also operated to the Isle of Man from Fleetwood. Apart from the GWR service from Rosslare all the Irish Sea services became part of the newly formed London Midland & Scottish Railway in 1923. Railway companies also operated passenger shipping services over the Severn, Humber and Clyde estuaries, across the Solent to the Isle of Wight, on lakes in the Lake District and on lochs in Scotland.

In Scotland the North British Railway (NBR) and the Glasgow & South Western Railway operated their own fleet of ships while the nominally independent Caledonian Steam Packet Company was formed in 1889 to operate steamer services for the Caledonian Railway along the Clyde and to the islands of the Clyde Estuary. After years of bitter competition between the rival shipping companies on the Clyde their fleets were amalgamated in 1923, becoming part of an enlarged Caledonian Steam Packet Company, the forerunner of today's Caledonian MacBrayne. At the same time the NBR's fleet of seven steamships, the Great Eastern Railway's twelve steamships and the Great Central Railway's thirteen steamships used on their North Sea services were absorbed by the newly formed London & North Eastern Railway in 1923.

There was also intense competition for the all important cross-Channel passenger ferry services from the South of England to France. By the early twentieth century three railway companies were fighting it out with a combined fleet of forty-eight steamships – the London & South Western Railway operated out of Southampton Docks, the London, Brighton & South Coast Railway out of Newhaven Harbour and the South Eastern & Chatham Railway out of Folkestone and Dover. All three railways and their fleets became part of the newly formed Southern Railway in 1923.

BELOW: The 'Big Four' railway companies' shipping services were also nationalized in 1948. Here the paddle steamer *PS Ryde* sailing under the Sealink flag crosses the Solent with a Portsmouth to Ryde service on 22 August 1965.

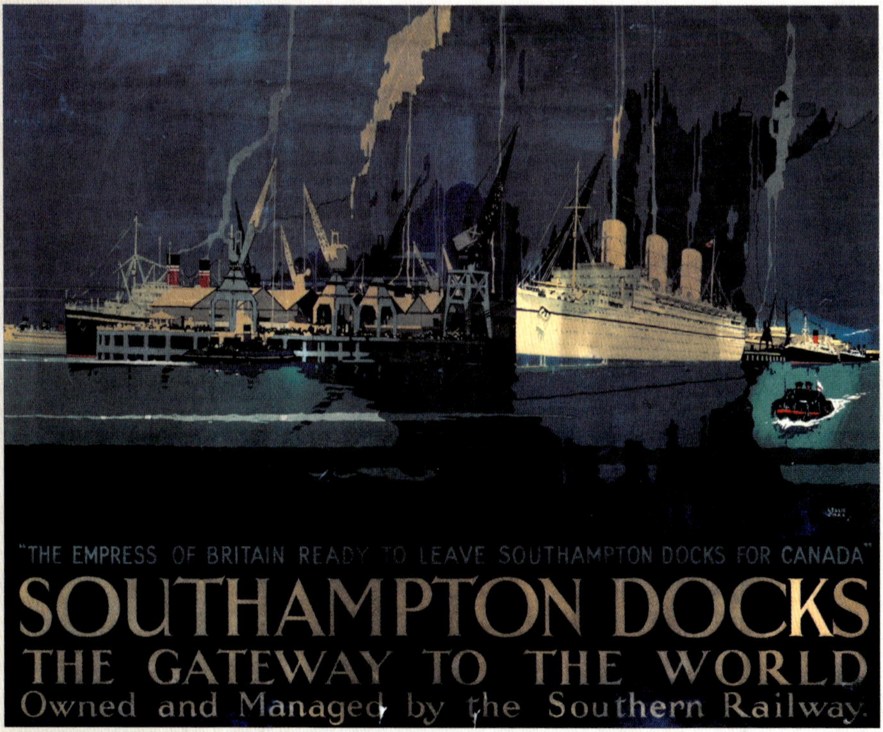

Freight traffic was even more important to railway companies than carrying passengers. By the early twentieth century many of the larger companies had built their own dock complexes to handle the vast amounts of imported raw materials and exported finished goods. At Southampton the London & South Western Railway (LSWR) had taken over the docks in 1892 and by 1911 had opened massive new facilities, which proved to be of vital strategic importance during the two World Wars, while in Lincolnshire the Great Central Railway opened its enormous dock facilities at Immingham in 1912. The LSWR's successor, the Southern Railway, continued the expansion of Southampton Docks in the 1930s making it one of the busiest ports in Europe.

Britain was also a major exporter of coal, all of it transported by rail from collieries to docks for onward shipment around the world. In the northeast the North Eastern Railway had a monopoly of this traffic and owned dock facilities at Hartlepool, Tyne Dock, Middlesbrough, Monkwearmouth and Hull as well as coal staithes at Blyth and Dunston-on-the-Tyne. In South Wales the Taff Vale Railway owned Penarth Docks, the Cardiff Railway owned Bute Docks and the most successful of them all, the Barry Railway, owned Barry Docks which at its peak in 1913 exported nine million tons of coal.

ABOVE: Southampton Docks was modernized and greatly enlarged by the Southern Railway in the 1930s. This SR poster produced in 1931 shows the brand-new *RMS Empress* of Britain about to depart from the docks for Canada. The ship was owned by the Canadian Pacific Steamship Company but was sunk by a German U-boat in 1940.

Joint railways

By the early twentieth century there were around 900 route miles of railways in Britain that were jointly owned by two or more railway companies. Some of these were strange bedfellows but in joining together all parties benefitted from the partnership in some form or another.

In Wales there were very few joint railways. The only two recorded were both jointly owned by the Great Western and the London & North Western railways: the 3-mile stretch of the GWR/LNWR line between the English/Welsh border and Buttington on the joint line between Shrewsbury and the Cambrian Main Line; and the 12½-mile section of the Central Wales Line between Llandeilo and Pontarddulais. Running just east of the English/Welsh border in the Welsh Marches, the 56-mile line between Shrewsbury and Hereford was also jointly owned by the same two companies.

Southern and southwest England had very few joint lines. One of the shortest in Britain could be found on the Isle of Wight, where the London & South Western Railway and the London, Brighton & South Coast Railway owned the 1¼-mile stretch between Ryde Pier Head and Ryde St John's Road. Probably the most famous of the joint railways was the much-loved 114-mile Somerset & Dorset Joint Railway between Bath and Bournemouth (and a branch to Highbridge) which, since 1891, had been jointly owned by the London & South Western and the Midland railways. In Gloucestershire the Great Western and the Midland railways owned the 39-mile Severn & Wye Joint Railway's route across the original Severn Bridge between Berkeley Road, Lydney and the Forest of Dean.

In the Home Counties the Great Western & Great Central Railways Joint Committee owned forty-one miles of track while in North London the Great Central also joined forces with the Metropolitan Railway to own and operate the line from Harrow-on-the-Hill to Verney Junction. Also in London a consortium of five companies owned the East London Railway.

The East of England had a fair share of joint railways, the longest (and the longest in Britain) being the 183-mile network of the Midland & Great Northern Joint Railway (M&GNJR) which stretched from Peterborough and Little Bytham in the west to Cromer, Norwich and Great Yarmouth in the east. In the same area the Norfolk & Suffolk Joint Railway owned two unconnected short stretches of line on the east Norfolk coast – the company was jointly owned by the Great Eastern Railway (GER) and the M&GNJR, the only instance of a joint railway part owning another joint railway! Eastern England also possessed the third longest joint railway in Britain – the 123-mile Great Northern & Great Eastern Joint Railway stretched from March to Doncaster along with another section south of March to St Ives and Huntingdon.

In the East Midlands the Great Northern & London & North Western Joint Railway was built primarily as a coal and iron ore-carrying line, at its northern end having connections with Nottingham, Newark and Grantham and in the south to Leicester and Market Harborough.

While the railways of the industrial northeast of England were all owned by one company, the North Eastern Railway, by far the largest concentration of joint

railways was found in South Yorkshire. Here an intricate network of railways, many of these jointly owned, competed for traffic in one of Britain's richest coalfields. The controlling interests of these joint lines ranged from the Lancashire & Yorkshire Railway (L&YR), the Great Northern Railway (GNR) and the North Eastern Railway (NER) to the Great Central Railway (GCR), the Midland Railway (MR) and the Hull & Barnsley Railway (H&BR). The two longest were the 27½-mile Axholme Joint Railway (NER and L&YR) and the 20-mile South Yorkshire Joint Railway (GCR, GNR, L&YR, MR and NER).

Over to the northwest the Cheshire Lines Committee operated 140 miles, making it the second longest joint railway in Britain. Owned jointly by the Great Northern Railway, the Great Central Railway and the Midland Railway it operated routes in Cheshire and Lancashire in competition with the London & North Western Railway. The CLC's main routes were Manchester Central to Liverpool Central, Manchester Central to Chester Northgate, and Liverpool Central to Southport Lord Street.

Other joint railways to be found in northwest England were the 9-mile Manchester South Junction & Altrincham Railway (LNWR and GCR), the 45-mile Preston & Wyre Joint Railway (L&YR and LNWR) and the 9½-mile Furness & Midland Joint Railway (Furness Railway and Midland Railway).

Finally to Scotland where, like Wales, there were very few joint railways. Probably the second shortest in Britain was the 1½-mile Forth Bridge Railway which was jointly owned by the North British, Great Northern, North Eastern and Midland railways. Down in the wilds of Galloway Scotland's only major joint railway was the 82-mile Portpatrick & Wigtownshire Joint Railway which was jointly owned by the Caledonian, Glasgow & South Western, London & North Western and Midland railways – strange bedfellows indeed!

LEFT: Somerset & Dorset Joint Railway 0-4-4T No. 54 halts at Wellow station with a Bath to Templecombe train in 1920. Much loved by local people and railway enthusiasts, the S&D was closed in 1966.

BELOW: Owned jointly by the Great Northern, Great Central and Midland railways, the Cheshire Lines Committee was the second longest joint railway in Britain. Its 140-mile network shown here in this 1920s' poster served Manchester, Liverpool, Chester and Southport.

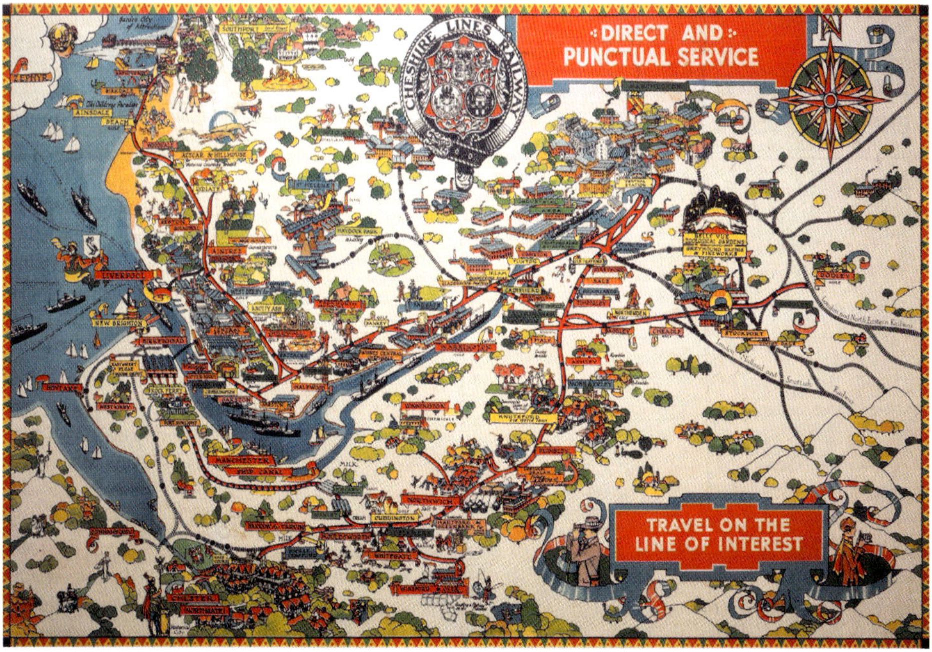

Island railways

Of approximately 6,000 islands around the coast of Britain most are just barren lumps of rock but there are around 800 of a significant size, of which 130 are inhabited. It is surprising to learn that quite a few of these had railways built on them, although the number still operating has declined significantly since the early twentieth century.

Channel Islands

There were once three railways on the Channel Islands but only one survives today. On Jersey the 7-mile Jersey Railway between St Helier and St Aubin, opening in 1870, was originally standard gauge. Later converted to 3-ft 6-in. gauge and extended to Corbiere, it closed after a serious fire at St Aubin in 1936. The 6¾-mile standard-gauge Jersey Eastern Railway opened between St Helier and Gorey in 1872 but was closed in 1929 due to road competition. A 2-mile standard-gauge line on Alderney was opened in 1847 transporting quarried stone to the new breakwater construction site at Braye. The one-time British Admiralty railway was taken over by preservationists in 1980 and currently operates a public service during summer weekends.

Wales

Trains running between Chester and Holyhead have served Anglesey and Holy islands in northwest Wales since the opening of Robert Stephenson's Britannia Bridge across the Menai Strait since 1850. There were also two branch lines on Anglesey: Gaerwen to Amlwch opened in 1864, closing to passengers in 1964 and freight in 1993; and the Red Wharf Bay branch opened in 1909, closing to passengers in 1930 and completely in 1950.

Isle of Man

The island once possessed a 46-mile network of 3-ft-gauge lines opened in the 1870s but today only the 15½-mile, steam-operated, section from Douglas to Port Erin remains open. Elsewhere on the island the 18-mile 3-ft-gauge electric tramway from Douglas to Ramsey opened at the end of the nineteenth century and is still working today. Another electric tramway, the 5-mile 3-ft 6-in.-gauge branch from Laxey to the summit of Snaefell, opened in 1895 and is also still working today. The 2-ft-gauge Groudle Glen Railway opened as a tourist attraction in 1896 and, after closing in 1962, was reopened in 1986.

England

The tiny Bristol Channel islands of Steepholm and Flatholm were taken over by the military in the Second World War to protect Bristol and Cardiff from air attack. Both islands had 60-cm-gauge tramways for resupply purposes. The track of the cable-operated switchback route from Steepholme jetty to the top of the island still survives. On Lundy Island a horse-drawn tramway trackbed from the 1860s can still be walked between the jetty and quarry.

In the South a horse-drawn tramway linking a pottery with a pier operated briefly on Brownsea Island in the 1850s while in the North East, on Lindisfarne, another horse-drawn tramway connected the lime kilns and jetty.

The Isle of Sheppey in Kent was first linked to the mainland by rail in 1860. Sheerness was reached via the Kingsferry Bridge from Sittingbourne by the London Chatham & Dover Railway and, now electrified, it is still operational. A light railway also opened in 1901 from the

main line at Queenborough to the coastal resort of Leysdown, closing in 1950.

The Hayling Island line from Havant to South Hayling opened in 1867 but was a victim of 'Dr Beeching's Axe', closing in 1963. The largest of England's island railways was on the Isle of Wight where by 1900 a network of fifty-six route miles connected the principal towns. Closures began in the 1950s, ending in 1966 when the remaining, steam-operated, routes between Ryde, Cowes and Ventnor closed. Ryde Pier Head to Shanklin was electrified using redundant third-rail London Underground trains a year later and is still operational. The Isle of Wight Steam Railway today operates the 5½-mile Wootton to Smallbrook Junction line.

Scotland

A network of narrow-gauge railways was proposed for the islands of Lewis and Skye in 1919 but none of these were built. The Isle of Skye did possess a 2-ft-gauge mineral railway between 1890 and 1915. Originally worked by manpower and gravity, the 2¾-mile Lealt Valley Diatomite Railway ran between diatomite drying nets on the shores of Loch Cuithir to a factory and jetty on the Sound of Raasay. In 1906 it was re-laid for steam operation but closed in 1915.

A 4-ft-gauge horse-drawn passenger tramway opened in 1882 on Bute in the Clyde Estuary. The roadside line between Rothesay and Port Bannatyne was rebuilt to 3-ft 6-in. gauge and electrified in 1902, being extended to Ettrick Bay in 1905. It closed in 1936.

Numerous short and unconnected narrow- and standard-gauge lines were built on the Orkney Islands from the early twentieth century to the Second World War, serving military installations, quarries, fishing harbours and a lighthouse service depot in Stromness Harbour. A 2-ft-gauge military railway network operated on Hoy from 1914 to 1920 during the construction of the Lyness Naval Base. Following the sinking of the 30,000-ton battleship 'HMS Royal Oak' by a U-boat in Scapa Flow in 1939, protection of this important naval anchorage was substantially increased with the construction of the Churchill Barriers. Contractors Balfour Beatty built a network of 2-ft and 3-ft-gauge lines, mainly on Burray and Hoy, used to transport enormous quantities of material required for construction of the concrete barriers.

LEFT: At its peak in the early twentieth century the railway network on the Isle of Wight reached fifty-six route miles. This poster was produced in 1910 for the Isle of Wight Railway and the Isle of Wight Central Railway to promote the island's attractions.

ABOVE: Class 'A1X' 0-6-0T No. 32661 crosses Langstone Harbour causeway with an evening Havant to Hayling Island train on 9 September 1960.

Railway road services

While early wagonways and tramways depended entirely on horsepower to keep them running, the successful introduction of steam power in the early nineteenth century virtually eliminated their use on all but a few railways. However, until the advent of the internal combustion engine horses still remained an important part of railway life well into the twentieth century.

Horsepower

Despite being superseded by steam locomotives to haul trains on railway lines, thousands of horses continued to be used by British railway companies for delivery of goods by road and for shunting in goods yards. By the early twentieth century the Midland Railway and the London & North Western Railway each owned over 5,000 horses, the Great Northern Railway and the Great Western Railway each owned nearly 3,000, and the Lancashire & Yorkshire Railway and the Great Eastern Railway each owned nearly 2,000. Shunting horses were normally employed at small country stations for moving single-wagon loads but they were prone to foot and leg injuries and were gradually replaced by powered capstans. Surprisingly the last railway shunting horses only retired from duty at Newmarket station in 1967.

By the early twentieth century the railway companies virtually monopolized the delivery of goods being carried to and from railway stations, goods yards and warehouses. Even by the Second World War, the majority of goods were still being delivered on horse-drawn wagons – at the time of the nationalization of railways in 1948 British Railways inherited over 8,000 horses. Horse-drawn delivery vehicles were eventually phased out in the late 1950s.

BELOW: Seen here in 1909, this GWR delivery van pulled by a 2-horse team operated from Paddington station, delivering parcels to the station for onward delivery in the guard's van of a passenger train.

RIGHT: Seen here at Torquay in 1927 this charabanc, or observation car, had glass panels at the side and a roof that could be rolled back in warm weather. They were used by the GWR as part of a day trip package for sightseeing tours.

Internal combustion engine

The introduction of motorized delivery vehicles was a long process. While small 3-wheeler delivery vans based on motorcycles started appearing in 1904, it took until 1932 when the famous Scammell Mechanical Horse made its debut – this consisted of a 3-wheeled tractor and an articulated trailer and its tight turning circle was perfect for working in the confined spaces of goods yards. The Scammell remained in production until the 1960s but by the end of that decade British Railways' road delivery services had virtually ceased to exist.

Meanwhile back in the early years of the twentieth century several railway companies started operating motorbus feeder services – serving sparsely populated country areas these were cheaper to operate than the cost of building a railway. The first company to run such a service was the narrow-gauge Lynton & Barnstaple Railway in North Devon, which introduced motor coaches between Blackmoor Gate station in Exmoor and the resort of Ilfracombe in 1903. However, this was not a success and after a short operating season the buses were sold to the Great Western Railway (GWR).

The GWR opened its first feeder bus service between Helston station and The Lizard in August 1903. It proved so popular that within a few years the company was operating routes in Devon, Somerset, the Cotswolds, the Thames Valley, the Midlands and Wales. By 1928 the GWR owned the largest railway-operated bus fleet in Britain but the legality of their monopoly was by then in question and in that year Parliament forced the transfer of their bus operations to nominally independent bus companies. However, these new companies were still part-owned by the GWR – in Devon and Cornwall the GWR had a 50 per cent stake in the Western National Omnibus Company.

Although not operating on such a grand scale as the GWR, other railway companies around Britain also introduced their own feeder bus services. Of note was the Great North of Scotland Railway (GNoSR) which, although it only owned a network of 333 route miles of railway, mainly in Aberdeenshire, had introduced its first feeder bus service between Ballater and Braemar in 1904. By the late 1920s the London & North Eastern Railway, successor to the GNoSR, operated fourteen bus routes in northeast Scotland with a total length of 290 route miles. However, by 1930 all these had been sold off to private companies.

Royal trains

The reign of Queen Victoria coincided with the railway revolution in Britain, and the monarch and her family were quick to realize that train travel offered speedier journeys in comfort and seclusion. By the time Victoria came to the throne in 1837 the railway network we know today was decades away – even the Great Western Railway (GWR) and the London & Birmingham Railway were yet to be completed.

The broad-gauge GWR was the first railway to receive royal patronage after opening between Paddington and Maidenhead in 1838. Slough, only a few miles from the royal residence of Windsor Castle, became the railhead for royal visits until 1849 when the GWR's and London & South Western Railway's (LSWR) separate branch lines opened to the town.

Royal patronage of the GWR began in 1839 when Prince Albert, Victoria's suitor, travelled between Paddington and Slough. Clearly impressed with this new mode of travel he persuaded Victoria to make a journey. The big day came on 13 June 1842 when the GWR rolled out the red carpet at Slough and conveyed the Queen and Prince Albert to Paddington. The GWR's top brass ensured a smooth journey, with Isambard Kingdom Brunel and Daniel Gooch riding on the footplate of brand-new broad-gauge 2-2-2 locomotive *Phlegethon*. This historic occasion was a success, with Victoria and Albert making the historic 25-minute journey in a specially fitted-out coach. Thus began an era of royal train travel that continues to this day. Not all royal trains were happy events as GWR funeral trains were also run from Paddington to Windsor following the death of a monarch – Queen Victoria's in 1901 was hauled by 4-4-0 No. 3373 (renamed *Royal Sovereign*), King Edward VII's in 1910, hauled by 4-6-0 No. 4021 *King Edward*, King George V's in 1936 behind No. 4082 *Windsor Castle* and King George VI's in 1952 behind No. 7013 *Bristol Castle* (renumbered and renamed as No. 4082 *Windsor Castle* for the day).

The King of France, Louis Philippe's state visit in 1844 was the first of many occasions when foreign royalty and dignitaries were taken by train to and from London – after landing at Portsmouth the King, along with Prince Albert and the prime minister, the Duke of Wellington, was conveyed from Gosport to Farnborough by the LSWR. The purchase of Osborne House on the Isle of Wight by the royal couple in 1845 led to a new royal station and pier being built at Gosport from where the royal yacht would cross the Solent to Cowes. It was from here that Queen Victoria's body was conveyed by royal train back to London following her death at Osborne House in 1901 – its use as a royal residence ended when the new king, Edward VII, gave it to the nation.

Royal train travel increased in 1848 when Victoria leased Balmoral House in the Highlands to the west of Aberdeen. The royal family normally spent their summers there and the royal train replaced the slow and often-uncomfortable sea journey in the royal yacht to Aberdeen. The opening of the Deeside Railway to Ballater in 1866 greatly improved the journey, with Victoria often making two visits each year – Ballater station's royal waiting room was last used by Queen Elizabeth II on 15 October 1965, just four months before the Deeside line closed.

In 1862 Sandringham in Norfolk was purchased as a country home for Prince Albert Edward – it was probably no coincidence that the nearby railway from King's Lynn to Hunstanton had also opened in that year. The nearest station to Sandringham was at Wolferton, on which Prince Edward later lavished much money. Sandringham became extremely popular with the royal family and between 1884 and 1911 no fewer than 645 royal trains used Wolferton station. Wolferton also played host to two royal funeral trains following the deaths of King George V at Sandringham in 1936 and King George VI in 1952.

Running a royal train was a major operational headache. Its itinerary was never publicized and to avoid delays a detailed timetable for each train was sent in advance to station masters and signalmen along the planned route. Stations en route were closed to the public and smartly dressed station staff stood to attention as the train passed through. A pilot engine travelled ahead of the train, passenger and goods trains were shunted out of the way, and the royal locomotive was polished like new.

From the very early days of royal trains, railway companies provided their distinguished passengers with sumptuous accommodation in luxury saloon coaches often fitted out by top London interior designers. The Prince of Wales (later Edward VII) made wide use of his own private saloons to visit races, shooting parties, lady friends and his beloved Sandringham. In 1864 the Great Eastern Railway built a 6-wheeled coach for his visits to Norfolk, as did the London, Brighton & South Coast Railway in 1877 for his South Coast trips. The latter company surpassed itself in 1897 by building an entire new train of American-style

Pullmans for the Prince and Princess of Wales. Fitted out in utmost luxury the royal saloon rode on a pair of six-wheel bogies. However, the GWR stole the show in 1897 when it built a complete 6-coach train for Queen Victoria to celebrate her Diamond Jubilee.

The dawn of the twentieth century and a new monarch saw the London & North Western Railway launch its new royal train, built at Wolverton Works. King Edward VII and Queen Alexandra could now travel around their kingdom in absolute luxury in individual 12-wheel saloons – of course the King's Saloon included the obligatory smoking compartment. This train was long-lived, used by successive monarchs with several of the semi-royal saloons remaining in use late into the 1960s. Not to be outdone, the Great Northern and North Eastern railways also introduced a new luxury royal train in 1908 for use on the East Coast Main Line to Balmoral.

During the Second World War the London Midland & Scottish Railway built two 56-ton all-steel 12-wheel saloons for King George VI and Queen Elizabeth, complete with armour plated window shutters.

The dawn of the Elizabethan age in 1953 saw BR operating three royal trains, all harking back to the golden age of Edwardian rail travel. Over the following two decades air-conditioned coaches based on standard BR rolling stock slowly replaced older vehicles. The most recent was fitted out at Wolverton Works in Buckinghamshire to celebrate Queen Elizabeth II's Silver Jubilee in 1977 – the train comprises eight refurbished BR Mk 3 coaches originally built in 1972 for the prototype High Speed Train and is hauled by claret-liveried Class 67 diesel-electric locomotives.

BELOW: Queen Victoria is seen riding north to Scotland in her royal train hauled by a Great Northern Railway locomotive in 1851. Royalty made this journey many times en route to Balmoral Castle, which was reached via the East Coast Main Line to Aberdeen and along Deeside to Ballater station.

Scottish mainland narrow-gauge railways

Despite proposals put forward in 1919 to build hundreds of miles of light railways in Scotland none of these ever saw the light of day. Unlike Wales, where many narrow-gauge lines were built in the nineteenth century to support the slate-mining industry, Scotland had very few.

Campbeltown & Machrihanish Light Railway

Located at the southern tip of the Mull of Kintyre, Scotland's only public narrow-gauge railway was completely isolated from the country's main rail network. The 2-ft 3-in.-gauge Campbeltown & Machrihanish Light Railway started life in 1876 as a steam-operated 4½-mile tramway connecting a colliery with the harbour at Campbeltown – it replaced a derelict canal that had opened back in 1794.

For nearly thirty years the colliery tramway led a fairly quiet life but in 1905 work started on rebuilding it as a passenger-carrying light railway, extending it westwards to a terminus behind the Ugdale Arms Hotel in Machrihanish. The new railway opened in August 1906 and was a great success, attracting thousands of passengers in its first month of operation – the Glaswegian day trippers arrived at Machrihanish on board fast steamers that had recently been introduced between the River Clyde and the Mull of Kintyre. The attractions of Machrihanish, with its golf links and beach, were not lost on the railway's publicity department. While still transporting coal the railway's success as a tourist attraction continued until the outbreak of the First World War. Although reintroduced after the war, the passenger service fell on hard times, facing ever-increasing competition from motorbuses. The closure of the colliery in 1929 sounded the death knell for the little railway, which closed for good in May 1932.

Cruden Bay Hotel Tramway

In 1897 the Great North of Scotland Railway (GNoSR) opened a 15-mile standard-gauge branch line from Ellon, on its Dyce to Peterhead line, to Boddam on Aberdeenshire's east coast. The line was primarily built to serve the GNoSR's new 55-bedroom luxury hotel at

Cruden Bay, which opened two years later. A 3-ft 6-in.-gauge electric tramway connecting Cruden station and the hotel opened in 1899 – the GNoSR built the two tramcars at its Kittybrewster Works in Aberdeen. Popular with golfers the hotel was initially successful but with falling passenger numbers the Boddam branch closed to passengers in 1932, hotel visitors being conveyed instead by motorbus from Aberdeen. However, the tramway continued to operate carrying laundry and supplies delivered to Cruden Bay station on the by-now freight-only line from Ellon. The tramway finally closed at the end of 1940 when the hotel was taken over by the military.

Shooting estate railways

Several grouse shooting estates in Scotland once operated narrow-gauge lines to transport shooting parties and their guns. Probably the largest was the 2-ft-gauge Duchal Moor Railway which was built by Sir James Lithgow in the Muirshiel Hills near Kilmacolm, Renfrewshire. Consisting of three connecting branch lines the 7-mile network was completed in 1922 and remained operational until the 1970s. Clutching their shotguns, cartridges, hampers and hip flasks, Sir James' guests (sometimes including royalty) were conveyed across the boggy moors in primitive coaches hauled by diminutive petrol-engine locomotives. The route of the railway can still be traced on modern Ordnance Survey maps.

Other 2-ft-gauge estate railways included the Dalmunzie Railway near Glenshee in east Perthshire, which was built in 1920, and the Ardkinglas Railway on the shores of Loch Fyne, which operated between 1879 and the early years of the twentieth century.

Lochaber Narrow Gauge Railway

While not a public railway, the most extensive narrow-gauge line to operate in Scotland was the Lochaber Narrow Gauge Railway which was built by Balfour Beatty in connection with the construction of a hydro-electric scheme to power a new aluminium smelter being built at Fort William. With no road access on the lower slopes of the Nevis range of mountains, the contractors built a 21-mile 3-ft-gauge railway to carry men and materials to numerous construction sites between Fort William, Loch Treig and Loch Laggan. Construction of this enormous civil engineering project started in 1925 and was only completed during the Second World War but the railway remained in use for maintenance work until the early 1970s. The trackbed can still be followed on foot today, while many of its American-style trestle bridges with track still in situ survive intact. A separate 1¾-mile 3-ft-gauge line that connected the aluminium smelter with jetties on Loch Linnhe survived until the early 1960s.

LEFT: Campbeltown in 1930 – passengers from a Clyde steamer wait to board the 'Steamer Express' to Machrihanish at the Hall Street terminus. The 2-ft 3-in.-gauge Campbeltown & Macrihanish Light Railway closed two years later.

ABOVE: Hunslet 0-4-ST 'Lady Morrison' of the 3-ft-gauge Lochaber Narrow Gauge Railway propels a load of cryolite from the pier on Loch Linnhe to the aluminium smelter at Fort William on 25 May 1961.

Railways at war (1)

With the clouds of war gathering over Europe, in 1914 the British Government handed the overall control of the country's railways to the Railway Executive Committee. Under the Committee each railway company was paid an amount based on their receipts for 1913. All non-essential traffic such as excursion trains was cancelled and railways struggled to cope with an increasingly rundown infrastructure, lack of maintenance and shortages in manpower – the latter caused by the thousands of railway workers that went to fight the war and never returned. To fill this gap many thousands of women found work on the railways but they were paid two-thirds less than the normal rate for men.

Britain's railway network was at times almost overwhelmed by the extra traffic generated by coal, munitions and troop trains – vast numbers of soldiers and their equipment were transported to ports in southern England to be embarked across the English Channel to France. Britain declared war on Germany on 4 August 1914 and within a few weeks special troop trains carrying members of the Expeditionary Force were leaving Waterloo, Liverpool Street and other stations for embarkation from the south coast ports. Southampton Docks was receiving a troop train every twelve minutes over a 14-hour day – by the end of the month London & South Western Railway trains had carried nearly 120,000 troops to Southampton along with 38,000 horses, 300 guns and over 5,000 military vehicles. Before long thousands of injured soldiers began returning in the reverse direction, carried in ambulance trains to hospitals around the country. In the early months of the war thousands of Belgian refugees arrived destitute at Harwich and Parkeston Quay and were transported to London by the Great Eastern Railway.

Enormous quantities of coal were delivered to the British Fleet based at Scapa Flow, with trains making the long overnight journeys from South Wales to the Far North of Scotland – these were nicknamed 'Jellicoe Specials' after the Admiral of the Grand Fleet, John Rushworth Jellicoe.

By the end of the war Britain's railway companies were in a poor physical and financial state so government control lasted until 1921. While there were calls for the nationalization of the railways, the 1921 Railways Act

dealt with the problem by grouping 120 railway companies into just four larger and more viable entities.

Railway Operating Division (ROD)

Across the English Channel the overloaded French railway system was falling apart. To deal with this the Railway Operating Division (ROD) of the Royal Engineers was formed in 1915. By the following year the ROD had requisitioned around 600 freight locomotives (mainly 0-6-0s) and thousands of goods wagons from British railway companies for use not only in France but also in Italy, Palestine, the Balkans and Mesopotamia. While a few of these locos were lost at sea most survived the war and were then either sold to the host country or returned home. With the war in France grinding to a standstill the demand for heavy freight locomotives far outstripped supply, so the ROD adopted the Great Central Railway's Class '8K' 2-8-0 as its standard freight locomotive. Over 500 of these powerful machines were built for service in Egypt, Palestine, Syria and Iraq, with the majority returning home at the end of the war.

War Department Light Railways

By 1916 the war in France had ground to a halt, with both sides dug in to hundreds of miles of waterlogged trenches in a blasted muddy landscape. Supplying the front line with supplies and munitions proved to be a difficult task as the early motor lorries got bogged down in the mud – although in short supply horses were the only answer for hauling artillery, ambulances and supply wagons. Faced with the problem of supplying the front line, the Ministry of Munitions recommended the use of narrow-gauge light railways that could be laid quickly and cheaply.

By 1917 the newly formed War Department Light Railways had built a network of 60-cm-gauge lines that at its peak extended to 700 miles and was capable of

LEFT: Women workers using turret lathes at the Lancashire & Yorkshire Railway's Horwich Works in 1917. Although paid less than men, thousands of women worked on the railways during the First World War, replacing men who had joined the army.

ABOVE: Lancashire & Yorkshire Railway Ambulance Train No. 24 – Britain's major railway companies all supplied ambulance trains during the war to transport injured soldiers from South Coast ports to hospitals around Britain.

supplying 7,000 tons of supplies and munitions each day. Steam locomotives built by various British and American companies, such as Hunslet and Baldwin, moved materials, supplies and munitions from standard-gauge railheads in the rear areas to marshalling yards nearer the front. From here small petrol-engine Simplex tractors that were less visible to the enemy completed the journey with their loads to the front line. Around 1,000 of these diminutive locomotives, some armoured, saw active service during the war.

RIGHT: Women were called to work on the railways during the war to replace men who had gone to Europe to fight. Here a female ticket inspector chats to an elderly railway worker on the Great Central Railway.

BELOW: During the war the Railway Operating Division (ROD) requisitioned hundreds of freight locomotives for use in northern France. Here, Great Eastern Railway Class 'Y14' 0-6-0 No. 534 is seen at work at Mordicourt on 9 October 1918.

Quintinshill Disaster

Quintinshill, ten miles north of Carlisle on the West Coast Main Line, was the scene of Britain's worst railway disaster on 22 May 1915. Due to errors by two signalmen a southbound fast-moving crowded troop train made up of fifteen old wooden carriages lit by acetylene gas ran head-on into a stationary local train which had stopped at the station. Although a major disaster in itself, much worse was to follow when minutes later a double-headed northbound express train ploughed into the wreckage at high speed. The accident site at Quintinshill then became a scene from Dante's Inferno, when the escaping acetylene gas from the wrecked wooden carriages of the troop train were ignited by red-hot coals from the derailed locomotives.

Of the 227 victims of this horrific crash all but twelve were soldiers of the Royal Scots Regiment on their way to fight the war in France. A further 245 were injured but the accident was hushed up due to wartime reporting restrictions. At their trial the two signalmen were found guilty of gross neglect of duty and culpable homicide, and were jailed.

ABOVE: The aftermath of the multiple collision at Quintinshill on 22 May 1915. In the ensuing inferno 227 people perished, most of whom were soldiers of the Royal Scots Regiment on their way to France.

Famous locomotive engineers

George Jackson Churchward

Born in 1857 in Stoke Gabriel, South Devon, George Jackson Churchward not only excelled at mathematics at school but he developed a love for the countryside which he retained for the rest of his life. At the age of 16 he was apprenticed at the South Devon Railway's (SDR) locomotive works at Newton Abbot where he trained as an engineer under the railway's Locomotive Superintendent, John Wright. The Great Western Railway (GWR) absorbed the SDR in 1876 and Churchward was transferred to the company's drawing office at Swindon Works. Within a year William Dean had taken over from Joseph Armstrong as the GWR's Locomotive Superintendent and in 1881 Churchward was promoted as Assistant Manager in the Carriage and Wagon Works, becoming Manager in 1885.

Churchward continued his meteoric rise when he was promoted to Locomotive Works Manager in 1895 and Principal Assistant to Dean in 1897, the same year that he was also elected as Mayor of Swindon. By this date Dean's health was failing and Churchward took over more and more of his responsibilities until 1902 when Dean retired and Churchward was promoted to the exalted position of Locomotive, Carriage & Wagon Superintendent. During his tenure at Swindon Churchward made great strides in reducing both construction and maintenance costs and introduced standardization of parts on a hitherto undreamed of scale. Using proven French and American practices, such as tapered boilers, he designed nine standard locomotive classes in the years between 1903 and 1911. In 1908 Churchward's driving force led to the building of the famous 'A' shop at Swindon Works – covering five acres it could produce seventy new locomotives and overhaul 600 each year.

Churchward's locomotive designs included the inside-cylinder 'City' Class 4-4-0s of which *City of Truro* became the first steam locomotive to achieve 100 mph in 1903. His legacy at Swindon was enormous with his powerful four-cylinder 'Star' Class 4-6-0s introduced in 1907 being the forerunner of the later 'Castle' Class locomotives designed by his successor C.B. Collett. Churchward's one failure was his unique *The Great Bear* which when built in 1908 was the first locomotive in Britain with the 4-6-2 'Pacific' wheel arrangement. It was not a great success as its weight limited it to operating only between London and Bristol. Succeeded by Charles Collett, Churchward retired in 1922 but was sadly killed in 1933 by one of his locomotives while inspecting track near his home in Swindon.

ABOVE: Built in 1908, Churchward's *The Great Bear* was the first 'Pacific' (4-6-2)-type locomotive to be built in Britain. It was not a great success as its route availability was limited due to its high axle loading and was rebuilt as 'Castle' Class 4-6-0 *Viscount Churchill* in 1924.

Dugald Drummond

Born in Ardrossan, Scotland, in 1840 Dugald Drummond was apprenticed at a Glasgow engineering company before being appointed as boiler shop manager at Thomas Brassey's works in Birkenhead. In 1864 he moved back to Glasgow where he worked under Samuel Johnson at the Edinburgh & Glasgow Railway's locomotive works in Cowlairs. Gaining further experience Drummond then worked under William Stroudley at the Highland Railway's Lochgorm Works in Inverness, moving with him to Brighton Works in 1870 to work for the London, Brighton & South Coast Railway.

Drummond left Brighton in 1875 to take the post of Locomotive Superintendent of the North British Railway at its Cowlairs Works in Glasgow. Here he designed seven classes of locomotive and was also called as an expert witness at the inquiry into the Tay Bridge disaster of 1879. Moving yet again in 1882, Drummond was appointed Locomotive Superintendent of the Caledonian Railway at its St Rollox Works in Glasgow where he designed nine classes of locomotive including the Class '294' 0-6-0 standard goods, many of which survived in service until the 1960s.

Drummond left the Caledonian Railway in 1890 to pursue his own engineering interests firstly in Australia and then back in Glasgow. Neither venture was very successful and in 1895 he took the post of Locomotive Engineer of the London & South Western Railway at its Nine Elms Works in London. During his 17-year tenure with the LSWR Drummond was a prolific locomotive designer, turning out no less than nineteen locomotive classes, many of which, such as the 'T9' Class 4-4-0 'Greyhounds' and 'M7' Class 0-4-4T, survived in operation until the early 1960s. Drummond died in harness in 1912 and is appropriately buried in Brookwood Cemetery, close to the former LSWR mainline in Surrey. Robert Urie succeeded him as CME of the LSWR.

Richard Maunsell

Richard Edward Lloyd Maunsell was born near Dublin in 1868 and after graduating from Trinity College in 1886 he was apprenticed at the Inchicore Works of the Great Southern & Western Railway (GS&WR) under Henry Ivatt. Following his apprenticeship Maunsell moved to England, where he worked in the Lancashire & Yorkshire Railway's locomotive works at Horwich before being promoted to locomotive superintendent at Blackpool. In 1894 he moved to India where he became a District Locomotive Superintendent for the East India Railway, returning to Ireland two years later where he eventually became Locomotive Superintendent of the GS&WR at Inchicore in 1911.

Maunsell only stayed at Inchicore for two years before he returned to England to take the post of Chief Mechanical Engineer (CME) of the South Eastern & Chatham Railway (SE&CR) at its Ashford Works. The SE&CR became a constituent company of the newly formed Southern Railway (SR) in 1923 and Maunsell was appointed its new CME, a post he held until retirement in 1937 when Oliver Bulleid succeeded him. Richard Maunsell died in 1944.

During his tenure at the SR Maunsell was responsible for producing eight new classes of locomotive, of which the most famous are the powerful 'Lord Nelson' Class 4-6-0 and the 'Schools' Class 4-4-0.

Maunsell's most powerful locomotives were the 'Lord Nelson' 4-6-0s. Here No. 850 *Lord Nelson* is seen on a boat train heading for Victoria station in 1928.

1923	1924	1925	1926	1927	1928	1929	1930	1931	1932	1933	1934	1935
'Big Four' Grouping			General Strike	Romney, Hythe & Dymchurch Railway opens		Swansea & Mumbles Railway electrified			GWR's *Cheltenham Flyer* becomes world's fastest train	Harry Beck's diagrammatic map of London Underground introduced	LNER 'A4' *Silver Link* achieves 112½ mph	LNE introduc *Silver Jubi* high-spee tra

THE 'BIG FOUR', STREAMLINERS and WAR

1923–1947

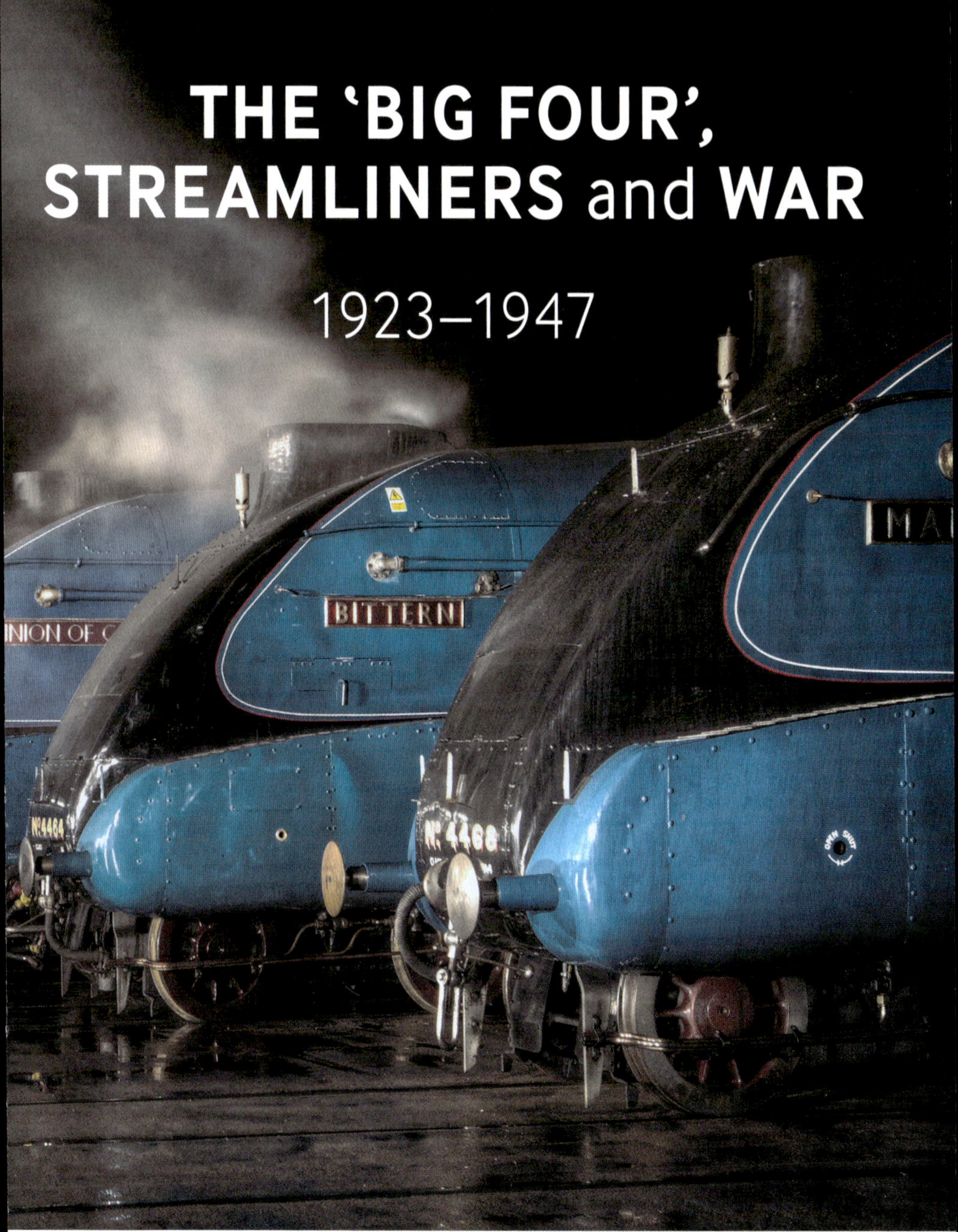

1936	1937	1938	1939	1940	1941	1942	1943	1944	1945	1946	1947
	LMS *Coronation* achieves 114 mph	LNER 'A4' *Mallard* sets unbroken world record of 126 mph	Wartime Railway Executive Committee take control of Britain's railways	320,000 troops evacuated from Dunkirk carried in 620 special trains		Ian Allan publishes his first ABC Guide for trainspotters		Railways move 2.5 million troops and their equipment for D-Day landings		LMS introduce first main line diesel electric locomotive	Transport Act

The 'Big Four' Grouping

Government control of Britain's railways during the First World War continued for nearly three more years after the end of the war. The war had brought Britain's 120 railway companies to their knees and they were in a bad shape both financially and physically. There was duplication of routes with many competing directly with each other, losses were mounting and the coalition government under the premiership of David Lloyd George wanted to introduce a more efficient and economical working of Britain's railway system. Some voices in Parliament called for nationalization and worker participation but this was rejected in favour of what became known as the 'Big Four' Grouping. Receiving Royal Assent on 19 August 1921, the 1921 Railways Act grouped the 120 railway companies into four larger regional companies, taking effect from 1 January 1923. The only exception was the amalgamation of London's railways, which was delayed until the 1933 London Passenger Transport Act.

Some railways were excluded from the grouping including eighteen standard-gauge light railways and fifteen narrow-gauge railways, many of which were managed by Colonel H.F. Stephens from his modest office in Tonbridge.

The 'Big Four' also inherited and further developed a network of feeder bus services but by the late 1920s the legality of these was in some doubt and the railway companies were forced to sell off their majority holdings in companies such as Crosville and the Bristol Tramways & Carriage Company. However, they did continue with their minority shareholdings in over thirty bus and coach companies.

The 'Big Four' inherited the multitude of shipping services operated previously by many of the constituent companies and also pursued other transport interests by forming airlines – the LMS, SR and GWR formed Railway Air Services (the predecessor to British European Airways) while the SR and GWR owned Channel Island Airways. Even the Thomas Cook travel company was sold to the 'Big Four'.

Great Western Railway

Of the 120 companies that made up the 'Big Four', the Great Western Railway (GWR) was the only one to retain its original name. With a total route mileage of 3,566 the newly enlarged GWR controlled large swathes of South and Mid-Wales, the West Country, the Cotswolds, the Thames Valley and the West Midlands. It had absorbed twenty-three other railway companies including the 295¼ route miles of the Cambrian Railways, the 124½ route miles of the Taff Vale Railway, the 63¼ route miles of the Midland & South Western Junction Railway and the 59¾ route miles of the Brecon & Merthyr Railway. The GWR also operated fifteen joint lines with the London Midland & Scottish Railway, two with the Southern Railway and one with the London & North Eastern Railway. The GWR also became the owner of three narrow-gauge lines in Wales: the Corris Railway; the Vale of Rheidol Railway; and the Welshpool & Llanfair Light Railway.

London Midland & Scottish Railway

With a network of 7,331 route miles the London Midland & Scottish Railway (LMS) was by far the largest of the 'Big Four'. It controlled the West Coast Main Line between

PREVIOUS SPREAD: No fewer than six of Sir Nigel Gresley's streamlined 'A4' 4-6-2s have been preserved, four in Britain and one each in Canada and the USA. With the latter two shipped over from North America, a unique gathering of all six took place at the National Railway Museum Shildon in 2014. This was probably the last time all locomotives will be together as the two visitors have since returned over the Atlantic to their respective museums.

LEFT: Ahead of the competition – this 1937 poster produced for the London & North Eastern Railway shows the company's achievement in introducing no less than four streamlined express trains.

BELOW: The 3,556-mile network of the enlarged Great Western Railway extended from Birkenhead in the north to Penzance in the south and from Fishguard in the west to London in the east.

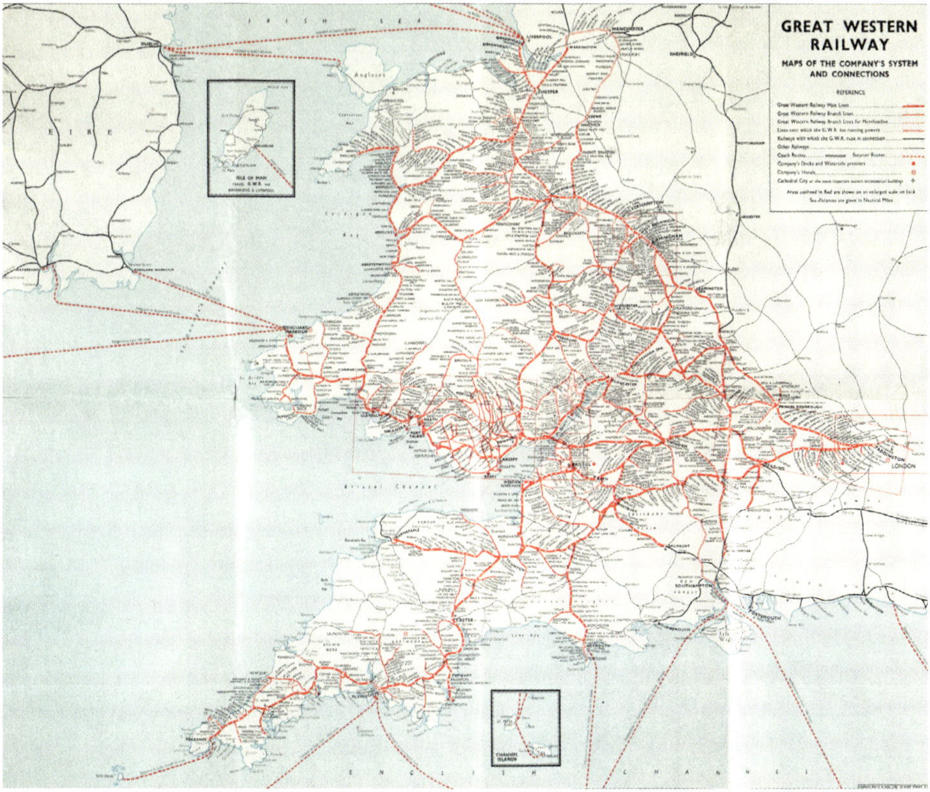

Euston and Glasgow and routes in North Wales, the Midlands, the North West of England and large swathes of Scotland extending up to Thurso and Wick. In England and Wales the LMS was formed by the amalgamation of the London & North Western Railway (including the Lancashire & Yorkshire Railway), the Midland Railway (including the Northern Counties Committee and the Dundalk, Newry & Greenore Railway in Northern Ireland and the London, Tilbury & Southend Railway), the North Staffordshire Railway and the Furness Railway. Scottish railway companies that became part of the LMS were the Caledonian Railway, the Highland Railway and the Glasgow & South Western Railway. The LMS also operated twenty-one joint lines (total mileage 585), of which the largest were the Midland & Great Northern Joint Railway in Norfolk and the Somerset & Dorset Joint Railway. It also became the owner of the narrow-gauge Leek & Manifold Valley Light Railway.

London & North Eastern Railway

The second largest of the 'Big Four', the London & North Eastern Railway (LNER) had a network of 6,671¾ route miles and controlled the East Coast Main Line from King's Cross to Edinburgh, East Anglia and Lincolnshire, the Great Central route from Marylebone to Sheffield, Yorkshire, North East England and Eastern Scotland. The major English constituents of the LNER were the Great Central Railway, the Great Eastern Railway, the Great Northern Railway, the Hull & Barnsley Railway and the North Eastern Railway. In Scotland the Great North of Scotland Railway and the North British Railway also became part of the LNER. The LNER also operated twenty-one joint lines with the LMS and one with the GWR.

Southern Railway

The smallest of the 'Big Four', the Southern Railway (SR) was the largest operator of passenger services with its busy commuter routes into South London. With a total network of 2,115½ route miles, the constituent companies that formed the SR were the London & South Western Railway, the South Eastern Railway, the London, Chatham & Dover Railway and the London, Brighton & South Coast Railway. The SR controlled railway routes in Southeast and Southern England, the Isle of Wight and Dorset, East and North Devon and North Cornwall. The SR's major joint line was the Somerset & Dorset Joint Railway between Bath and Bournemouth that it shared with the LMS. It also became the owner of the narrow-gauge Lynton & Barnstaple Railway.

BELOW: The London Midland & Scottish Railway's only streamlined train, 'The Coronation Scot', storms out of London Euston on the first part of its journey to Glasgow Central on 27 June 1938.

TOP RIGHT: Seen in this 1937 LMS poster, 'The Royal Scot' was the company's premier express train between London and Glasgow until the streamlined *Coronation Scot* was introduced in 1937.

BOTTOM RIGHT: Although the smallest of the 'Big Four', the Southern Railway operated the most passenger services with its busy commuter routes into South London. On its formation in 1923 the company's 2,115½-mile network stretched from Ramsgate in the east to Padstow in the west.

LMS
6110
Bryan de Grineau
SCOT PASSES SCOT
LMS
By Bryan de Grineau.
10·0 A.M. EUSTON TO GLASGOW AND EDINBURGH
10·0 A.M. GLASGOW AND EDINBURGH TO EUSTON
LONDON MIDLAND & SCOTTISH RAILWAY COMPANY

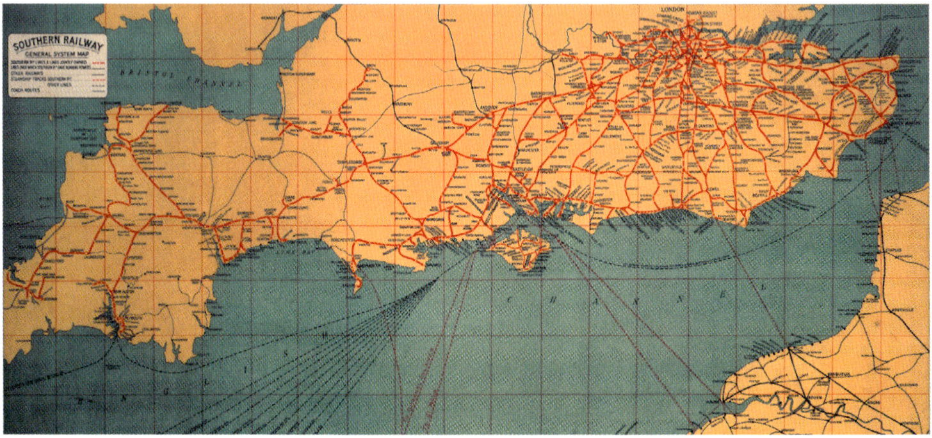
SOUTHERN RAILWAY
GENERAL SYSTEM MAP
COACH ROUTES
LONDON

English narrow-gauge railways

Unlike Wales where many narrow-gauge railways had been built to support the slate quarrying industry, very few were built in England. The majority of those that were built in the nineteenth century were fairly modest affairs supporting local industries, quarries or mines and were short lived. The 1896 Light Railways Act breathed new life into the English narrow-gauge scene, allowing companies to build new railways in sparsely populated regions of the country without expensive legislation and to much less exacting standards than previously required. Surplus equipment and track that became available from the War Department Light Railways after the First World War breathed new life into existing narrow-gauge railways and even created several new enterprises. However, all of the English narrow-gauge railways had fairly short lives and all had disappeared by the 1950s.

BELOW: Opening in 1879, the 3-ft-gauge Southwold Railway in Suffolk operated a service of mixed trains between the port of Southwold – seen here in the early twentieth century – and the town of Halesworth in Suffolk for fifty years.

RIGHT: Opened in 1898, the 1-ft 11½-in.-gauge Lynton & Barnstaple Railway served scattered farming communities in North Devon. It was taken over by the Southern Railway in 1923 and, despite modernization, was closed in 1935.

Snailbeach District Railway

Pontesbury to Snailbeach

Gauge: 2 ft 4 in. *Opening date:* 1877

This 3½-mile line was originally opened to transport lead ore from mines in the Stipperstones Hill in Shropshire to the GWR's standard-gauge railhead on the Minsterley branch at Pontesbury. Coal from Pontesford returned in the opposite direction. The railway was initially successful with ore traffic peaking at just under 40,000 tons in 1909, but from then on decline set in and by the 1920s it had almost ceased operation. Then in 1923 the doyen of light railways, Colonel H.F. Stephens, stepped in, purchasing the railway and re-equipping it with surplus War Department locomotives. During the 1930s the railway prospered transporting road stone from a new quarry and in 1947 was leased to Shropshire County Council who continued to use part of it until complete closure in 1959.

Southwold Railway

Halesworth to Southwold

Gauge: 3 ft *Opening date:* 1879

In Suffolk the charming 8¾-mile Southwold Railway followed the River Blyth between Halesworth, where it met the Great Eastern Railway's East Suffolk Line, and the

small harbour town of Southwold. The railway operated services with diminutive steam locomotives hauling mixed trains of both freight and passengers – the latter were carried in coaches unusually fitted with three bogies making the ride less than comfortable. The only major engineering feature on the line was a swing bridge across the Blyth near Southwold. Plans to convert the line to standard gauge never materialized and the railway closed in 1929. There are currently plans to reopen a short section near Southwold.

Rye & Camber Tramway
Rye to Camber Sands
Gauge: 3 ft *Opening date:* 1895
Engineered by Colonel H. F. Stephens this 1¾-mile line was built on private land through sand dunes from the Sussex town of Rye to Rye Golf Club and Camber Sands. From opening day on 13 July 1895 this little railway, which initially only operated one passenger coach hauled by a small 2-4-0 steam locomotive, went on to carry nearly 20,000 golfers, fishermen and day trippers in its first six months of operation. It was later extended to a new terminus in the sand dunes at Camber Sands but while passenger receipts held up well during the summer months the railway struggled to make ends meet in the winter, relying on subsidies from the golf club to keep it going. Closure came at the onset of the Second World War in September 1939 although the Admiralty used a section of the line for a while. The route of this delightfully quirky narrow-gauge line can be easily traced today.

Lynton & Barnstaple Railway
Barnstaple Town to Lynton
Gauge: 1 ft 11½ in. *Opening date:* 1898
Serving the up-and-coming coastal resorts of Lynton and Lynmouth and with financial backing from publisher Sir George Newnes, the highly scenic 19-mile Lynton & Barnstaple Railway opened to great celebrations in North Devon. Three handsome 2-6-2 tank locomotives and seventeen bogie coaches provided services between Lynton station and Barnstaple Town station where the L&BR met the London & South Western railway's standard-gauge branch line to Ilfracombe. However, many stations were situated some distance from the

places they purported to serve and journey times were inexorably slow. Facing increasing competition from motorbuses, the line was taken over by the newly formed Southern Railway in 1923 which promptly set about modernizing it. Sadly, in the end, this was all to no avail and the L&BR closed in 1935. Fortunately a section of the line from Woody Bay station has been reopened in recent years by preservationists while soaring Chelfham Viaduct still stands in all its glory.

Leek & Manifold Valley Light Railway

Waterhouses to Hulme End

Gauge: 2 ft 6 in. *Opening date:* 1903

Built under a Light Railway Order and worked by the North Staffordshire Railway (NSR), this 8-mile line once ran along the pretty Manifold Valley in Staffordshire. Depending more or less entirely on summer day trippers and milk traffic for its livelihood, the Leek & Manifold made a connection with the NSR's standard-gauge terminus at Waterhouses. The railway became part of the newly formed London Midland & Scottish Railway (LMS) in 1923 but with ever-increasing competition from road transport it closed in 1934. The LMS donated the trackbed to Staffordshire County Council and it was reopened as a footpath and bridleway in 1937. Now known as the Manifold Way it is today a popular destination for walkers and cyclists.

Ashover Light Railway

Clay Cross to Ashover

Gauge: 1 ft 11½ in. *Opening date:* 1924

The last new major narrow-gauge line to open in England, the 7-mile Ashover Light Railway had its beginnings in George Stephenson's Clay Cross Company which had manufactured iron at Clay Cross in Derbyshire since the mid-nineteenth century. Surveyed by Colonel H.F. Stephens, the railway was built under a Light Railway Order using surplus track and locomotives bought cheaply from the War Department. While its primary use was to transport mineral deposits – limestone, fluorspar and barytes had been discovered in nearby Amber Valley in 1918 – to the Clay Cross Ironworks, the railway benefitted greatly from day trippers visiting the scenic Amber Valley during the summer months. All went well until 1936 when passenger trains were withdrawn in the face of increasing bus competition. The line struggled on and in its last years transported ballast for British Railways from a local quarry. Completely rundown and in a state of disrepair, England's last narrow-gauge railway finally closed in 1950.

BELOW: Built with war surplus equipment, the 7-mile Ashover Light Railway was the last new major narrow-gauge line to open in Britain. Serving mineral quarries in the Amber Valley in Derbyshire, it closed to passengers in 1936 and completely in 1950.

Miniature railways

Decades before the trainspotting craze hit Britain, Eton-educated Sir Arthur Heywood was a true Victorian railway enthusiast. As a teenager he had built himself a model locomotive and later, as one of the landed gentry, went on to develop a working 15-in.-gauge railway around the grounds of his home at Duffield Bank near Derby. His pioneering 'minimum-gauge' railway attracted great interest especially at it had possible military potential, and he used it as a test-bed for his home-built locomotives and rolling stock until his death in 1916.

Heywood's 'minimum-gauge' railway attracted much interest from the Duke of Westminster who commissioned him to build a working estate railway at Eaton Hall in Cheshire. Opened in 1896 and remaining in operation until 1946, the Eaton Hall Railway was an extensive affair with a network of 4½ miles of line connecting the hall, estate works and brickyard with a coal exchange siding at the GWR station at Balderton.

Another 'minimum-gauge' estate railway was the 5¼-mile 18-in.-gauge Sand Hutton Light Railway in East Yorkshire, which operated on the estate of Sir Robert Walker. Opening in 1922 it replaced a shorter 15-in.-gauge line and was built with and operated by surplus War Department equipment. Linking the estate village of Bossall and brickworks at Claxton with the LNER exchange siding at Warthill, the railway transported coal, bricks and agricultural produce and for a short time also operated passenger trains. It closed in 1932.

Meanwhile, up on the Cumbrian Coast the 3-ft-gauge Ravenglass & Eskdale Railway had opened in 1875 to transport hematite iron ore from mines in Eskdale down to the Furness Railway's transfer sidings at Ravenglass. Although this was England's first narrow-gauge line it only survived until closure in 1913. Two years later it was bought by model engineers W.J. Bassett-Lowke and R. Proctor-Mitchell who converted the line to 15-in. gauge, reopening it in 1917. Serving a granite quarry and the scattered communities along Eskdale, the little railway continued to operate until the end of the Second World War when the Keswick Granite Company bought it. The granite quarry closed in 1953 and the railway was taken over by a group of preservationists in 1960, which continues to operate it to this day.

On the West Wales coast near Barmouth a 2-ft-gauge horse-drawn tramway had opened between Fairbourne and Penrhyn Point in 1895. In 1916 it was converted to

15-in. gauge by W.J. Bassett-Lowke with passengers being carried in open-top carriages hauled by one of Henry Greenly's locomotives. Apart from a break between 1940 and 1947 the railway continued operating under a succession of owners until 1984 when it was converted to 12¼-in. gauge. Despite recent doubts over its future it still operates steam-hauled trains today.

Building miniature railways was a rich man's hobby and in the 1920s two wealthy men, Captain Jack Howey and racing driver Count Zborowski, planned to realize their dreams by opening the world's smallest public railway along the Kent coast between Hythe and Dymchurch. Sadly Count Zborowski was killed in a motor racing accident before the project had got off the ground so Captain Howey teamed up with miniature steam locomotive engineer Henry Greenly. Opening in 1927, the 15-in.-gauge 8-mile double-track Romney, Hythe & Dymchurch Railway is a one-third scale working miniature of a main-line railway complete with powerful Pacific-type steam locomotives. It was extended a further 5½ miles to Dungeness in 1928 and soon became a big hit with holidaymakers during summer months. During the Second World War the railway operated an armoured train fitted with machine guns and also carried materials for Operation Pluto. Howey's death in 1963 brought uncertainty over the railway's future but after several brief ownerships it was saved in 1973 and still operates along the Kent coast today.

Miniature railways also proved very popular attractions at many of Britain's seaside resorts from Edwardian times until the 1960s. A few of these have somehow survived and even use some of their original locomotives and rolling stock. Of note is the 15-in.-gauge Rhyl Miniature Railway in North Wales, which opened around a boating lake in 1911 and still uses steam locomotives designed by Henry Greenly and built by Albert Barnes in the early 1920s. In northeast England the 20-in.-gauge North Bay Railway in Scarborough has been running since 1931 and the 15-in.-gauge Saltburn Miniature Railway since 1947, while in Lincolnshire the 15-in.-gauge Cleethorpes Coast Light Railway has been giving enjoyment for nearly seventy years. Up in Scotland the 10¼-in.-gauge Kerr's Miniature Railway at Arbroath is still going strong since opening in 1935.

LEFT: The world's smallest public railway – the 15-in.-gauge Romney, Hythe & Dymchurch Railway station at New Romney. On the left is 4-8-2 No. 6 *Samson* and on the right 4-6-2 No. 10 *Dr Syn*, the latter fitted with a whistle from an LNER 'A4'.

BELOW: Originally built as a 3-ft-gauge line, the Ravenglass & Eskdale Railway in Cumbria was rebuilt to the 15-in.-gauge in 1917. Here miniature 2-8-0 locomotive *Ella* hauls a train of iron ore in 1927.

The London Underground

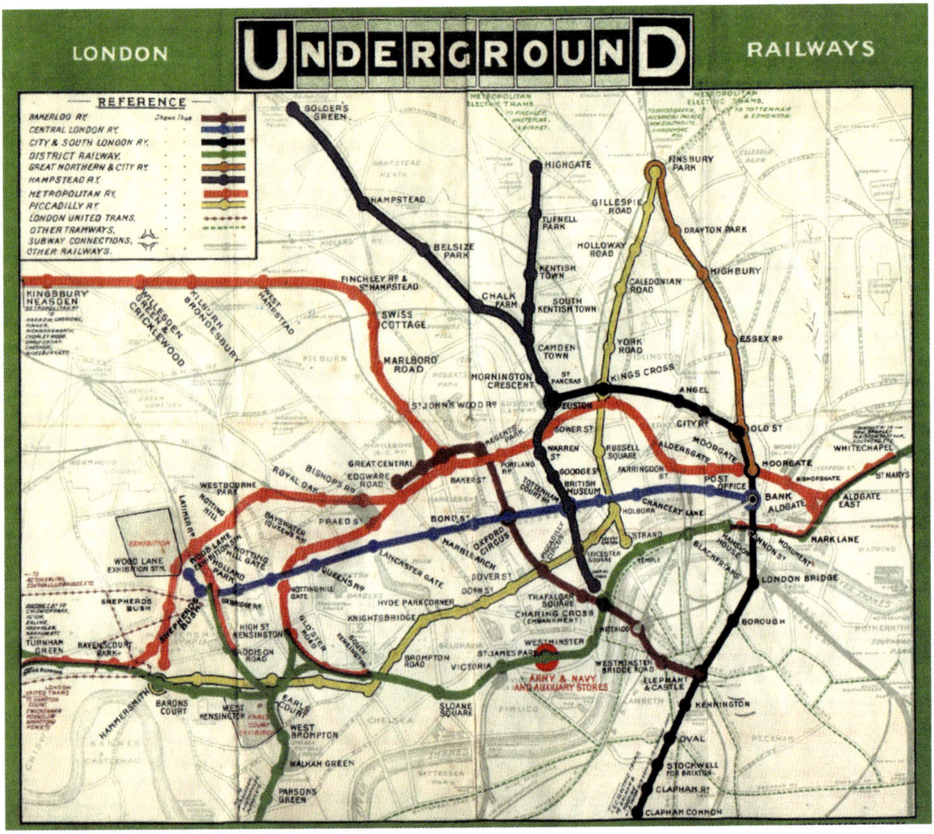

The early years of the twentieth century had seen enormous progress on London's underground railway network. By 1908 much of today's system was already in place and the three electric deep-level tube railways – the Baker Street & Waterloo, the Great Northern, Piccadilly & Brompton and the Charing Cross, Euston & Hampstead – were all controlled by the Underground Electric Railways Company of London (UERL). Founded by American financier Charles Yerkes in 1902, the UERL also owned the District Railway which had been electrified by 1905. Power for the network was generated at Lots Road Power Station alongside Chelsea Creek, which became operational in 1905 and consumed 700 tons of coal every day. It was later converted to run on oil and then natural gas before closing in 2002.

The world-famous roundel synonymous with London Underground was first used in 1908 when the UERL used solid red circles to highlight station names. In 1915 this design evolved into the red ring that is still used today and which was registered as a trademark in 1917. The sans-serif typeface that is still used today on all London Underground signs was created for UERL by Edward Johnston and first used in 1916.

While the onset of the First World War delayed extensions to the deep-tube network, government-backed financial guarantees were used after the war to expand the UERL network under central London. In the 1920s there were extensions to the City & South London and the Charing Cross, Euston & Hampstead Railway with what later became known as the Northern Line being completed in 1926. New Standard Stock trains with air-

LEFT: Drawn to scale, this 1908 map of the London Underground concentrates only on Central London, omitting the extremities of the District and Metropolitan lines. Still the basis of today's London Underground maps, Harry Beck's famous diagrammatic map of 1933 shows the whole network.

ABOVE: During the 1930s the Piccadilly Line was gradually extended northwards, reaching Arnos Grove in 1932 and Cockfosters in 1933. Here a tube train enters Bounds Green station on opening day, 12 December 1928.

operated sliding doors were introduced and lifts at busy central London stations were replaced with escalators.

The post-war years saw the Metropolitan Railway ('Met') develop much of its land in the northern leafy suburbs for housing estates – the term 'Metro-land' was first used by the 'Met's' publicity department in 1915 when the company was promoting a dream of modern homes built in the countryside served by fast commuter trains to central London. With government backing the 'Met' also extended its electrified line from Harrow to Rickmansworth in 1925.

Government-backed expansion continued into the early 1930s. The 'Met' opened a line from Wembley Park to Stanmore and the Picccadilly Line had reached Cockfosters in the north and Uxbridge in the west by 1933. This period was also marked by the building of Modernist-style station buildings designed by Charles Holden, the architect of the UERL's headquarters at 55 Broadway that had been completed in 1929. Constructed of reinforced concrete and glass, eight of his striking stations were built for the Piccadilly Line northern extension from Finsbury Park to Cockfosters – their historical importance today is reflected in their current Grade II listed building status.

The first integrated map of the London Underground network was published by UERL in 1908. It wasn't perfect because it was drawn to scale and therefore omitted the extremities of both the District and Metropolitan lines. Various geographic representations followed until the first diagrammatic, or topological, map was published in 1933. The map was designed for UERL by Harry Beck in his spare time while working as a draughtsman for the London Underground Signals Office and the style is still used today to illustrate London's Underground network.

While Britain's 110 railway companies had been grouped as the 'Big Four' in 1923 public transport in London remained privately owned until 1933. On 1 July all the UERL's deep-tube lines, the Metropolitan Railway and tram and bus companies were merged into one public corporation, the London Passenger Transport Board (LPTB). Two years later the government-backed £42 million New Works Programme (NWP) was launched to develop public transport services run by the LPTB and suburban services operated by the GWR and LNER. Designed for completion by 1940, the NWP got off to a good start before the Second World War brought a halt to proceedings. By 1940 the Northern Line had been extended to High Barnet although an extension beyond Edgware was halted as was electrification of the former GNR branch from Finsbury Park to Alexandra Palace. The Bakerloo Line had been extended to Stanmore but eastward extension of the Central Line remained unfinished until after the war. One of the most important improvements was the introduction of standard modern rolling stock for the deep-tube lines in 1938 – fitted with air-operated sliding doors these red tube trains remained in service in London for fifty years while several still operate on the Isle of Wight Island Line between Ryde Pier Head and Shanklin.

During the Second World War seventy-nine of London's deep-tube stations became air-raid shelters where thousands of Londoners would escape Luftwaffe bombing during the Blitz. The stations were fitted with bunks, special sanitary equipment and clinics along with 124 canteen points to which eleven tons of food was delivered daily by special train. During the early days of the Blitz 177,000 civilians took shelter each night on the Underground. However, their safety was not always assured as direct hits on Bank, Balham and Trafalgar Square stations proved.

The quest for speed

Ever since the *Rocket* achieved just over 29 mph at the Rainhill Trials in 1829 Britain's locomotive engineers continually strived to improve performance. It was on Brunel's broad-gauge Great Western Railway that speed records were set in the 1840s – with a pair of 8-ft-diameter driving wheels the 2-2-2 *Great Western* must have made quite a sight when it reached over 74 mph at Wootton Bassett on 1 June 1846. The broad gauge continued in the tradition when the Bristol & Exeter Railway's 4-2-4 tank locomotive No. 41, fitted with 9 ft driving wheels, reached nearly 82 mph down Wellington Bank in Somerset in 1854. The first locomotive to achieve 90 mph was the Midland Railway's 2-4-0 No. 117 in 1897 while travelling between Melton Mowbray and Nottingham.

The last decade of the nineteenth century was marked by the 'Railway Races to the North' when trains on the East Coast and West Coast main lines competed with each other for the fastest journey time between London and Aberdeen. A high-speed derailment at Preston put an end to the racing and strict speed limits were then enforced on both routes, which remained in place until the 1930s.

The magic 100 mph mark was allegedly reached in 1904 when the GWR's 4-4-0 No. 3717 *City of Truro* was hauling an 'Ocean Mails' special from Plymouth to Paddington, reaching the speed as it descended Wellington Bank. Locomotive speed records then took a break with the austerity years of the First World War and the 1920s putting a stop to such antics.

Aviation was still in its infancy by the 1930s and railway companies across the world were striving for ever-faster services. The 'Big Four' railway companies in Britain were getting their act together with the development of more powerful steam locomotives and modern coaching stock. The rivalry was intense between these publicity-seeking companies.

The Great Western Railway, always at the forefront of railway technology, was first off the starting block with its world-beating 'Cheltenham Flyer'. This famous train, hauled by Collett's new 'Castle' Class 4-6-0 locomotives, gained the title 'The World's Fastest Train' in 1932 when *Tregenna Castle* and its six coaches took just under fifty-seven minutes to travel the 77¼ miles between Swindon

BELOW: Headed by 4-6-0 No. 5000 *Launceston Castle*, the GWR's 'Cheltenham Flyer' crosses the Thames at Maidenhead. It became the world's fastest regular train in 1932.

TOP RIGHT: Seen here heading 'The Coronation Scot' in 1937, the LMS's streamlined 4-6-2 No. 6220 *Coronation* achieved 114 mph on a test run south of Crewe in June of that year.

BOTTOM RIGHT: Sitting on the turntable at King's Cross, the LNER's 'A4' 4-6-2 No. 2509 *Silver Link* was the first of its class when it achieved 112.5 mph in 1935. Along with three other similar locomotives it hauled the high-speed 'Silver Jubilee' express between London and Newcastle.

SILVER LINK
L

and Paddington at an average speed of 81.6 mph. The train soon officially became the fastest scheduled service in the world, regularly taking sixty-five minutes for the journey.

Elsewhere in the world the American 'Burlington Zephyr' and the German 'Flying Hamburger' were setting new standards using streamlined high-speed diesels. After visiting Germany, the LNER's Chief Mechanical Engineer, Nigel Gresley, carried out trials with his 'A3' Pacific *Papyrus*, setting a speed record of 108 mph down Stoke Bank on the East Coast Main Line and proving that steam was every bit as good as diesel. This success led to Gresley designing his world-famous 'A4' streamlined 3-cylinder Pacifics, the first of which, *Silver Link*, reached a speed of 112.5 mph, also down Stoke Bank, while hauling a rake of seven streamlined articulated coaches in September 1935. The new 'A4s' were soon in regular service hauling the 'Silver Jubilee' express between King's Cross and Newcastle, the locos proving to be very reliable and popular with their crews.

The LMS, not to be outdone by the LNER's success, introduced its 'The Coronation Scot' streamlined train between Euston and Glasgow in 1937. Motive power was provided by William Stanier's new 'Coronation' Class Pacifics, the first ten members being fitted with streamlined casings, hauling a rake of matching coaches. During a test run in June 1937, No. 6220 'Coronation' achieved a speed of 114 mph just south of Crewe, narrowly avoiding a derailment as it approached the station at an excessive speed.

BELOW: Gresley's 'A4' 4-6-2 No. 4489 *Dominion of Canada* leaves King's Cross station with the trial run of the streamlined 'Coronation' high-speed express. Introduced in July 1937, the train called at York and Newcastle before arriving in Edinburgh in exactly six hours.

Mallard's record-breaking run

The LNER's 'A4' Pacific No. 4468 *Mallard* was fresh out of Doncaster Works on 3 March 1938 and after running in was prepared to break the world speed record for steam locomotives of 124½ mph, set by German streamlined locomotive No. 05 002 in 1936. On 3 July 1938 the 165-ton *Mallard* set off southwards from Grantham on the East Coast Main Line with driver Joseph Duddington in control aided by his fireman Thomas Bray. At the rear were six coaches and a dynamometer car totalling 240 tons. *Mallard* topped Stoke Summit at 75 mph and on the long descent to Essendine kept piling on the speed until 126 mph was recorded as the train flashed past Milepost 90¼ – the world record had been broken and it still stands to this day.

These high jinks were halted in 1939 with the onset of the Second World War and this Golden Age was never to return in the post-war austerity years. The LNER 'A4's were shorn of their streamlined side valances and the LMS 'Coronations' lost their streamlined casings, although the 'Elizabethan' non-stop express hauled by 'A4's between King's Cross and Edinburgh provided a little glamour until it was withdrawn in 1961.

Fortunately six members of the 'A4' Class (including *Mallard*) have been preserved, including one each in museums in the USA and Canada. In 2014 the latter were shipped over to the UK to take their place in the 'Great Gathering' of 'A4's at the National Railway Museum in York before returning back over the Atlantic to their respective homes. The preserved ex-LMS 'Coronation' Class Pacific *Duchess of Hamilton* has recently had its streamlined casing reinstated, and the loco can be seen in all its glory at the National Railway Museum in York.

ABOVE: Gresley's 'A4' 4-6-2 No. 4468 *Mallard* achieved a world speed record for steam locomotives on Sunday 3 July 1938. The locomotive reached a recorded 126 mph south of Grantham on the East Coast Main Line, a record that it still holds today.

Minimal trains

Railmotors

The early years of the twentieth century saw several British railway companies introduce steam railmotors with mixed success. These single-carriage trains were theoretically cheap to build, economical to operate and were ideal on routes where passenger numbers were light, but in reality they were incapable of hauling extra rolling stock during busy periods and still needed a driver, fireman and guard to operate them. By far the largest user of steam railmotors was the Great Western Railway (GWR), which first introduced them between Chalford and Stonehouse in Gloucestershire in 1903. Designed by G.J. Churchward the new railmotors were fitted with internal vertical boilers and Walschaerts valve gear (unusual for the GWR), and could be driven from either end. With a top speed of 45 mph, tickets issued by the guard and seating fifty-two passengers they were a great success, serving numerous small halts along the 7-mile route along the Golden Valley. The GWR went on to build ninety-nine of these railmotors, many of which saw service in cities such as Plymouth and on short routes out of Paddington where they competed with electric street trams for business. The last one was withdrawn in 1935 by which time their operations had been taken over by push-pull trains (see below).

Other railways to use railmotors with mixed success included the Taff Vale Railway, Furness Railway, Great Central Railway, Great Northern Railway, the London & South Western Railway and the Lancashire & Yorkshire Railway (L&YR). Unlike the GWR's railmotors the L&YR's version consisted of a diminutive 0-4-0 tank locomotive which was permanently coupled to a single coach. The latter was supported at one end by the locomotive and at the other by a 4-wheel bogie – this self-contained steam train could be driven from either end and required no turning. Introduced in 1911 a total of twenty were built for use on branch lines in Lancashire with the last remaining in service until 1948.

The final evolution of steam railmotors came in 1924 when the Sentinel Waggon Works of Shrewsbury exhibited its latest model at the 1924 British Empire Exhibition. The London & North Eastern Railway (LNER) ordered eighty of these self-contained passenger-carrying vehicles, which were delivered between 1925 and 1932. Some were fitted with vertical boilers while others had horizontal boilers and were powerful enough to haul a trailer car at speeds of up to 60 mph. They could be seen at work on branch lines and short suburban routes across the LNER network, their box-like appearance and smoking chimney giving them the nickname of 'chip vans'. Their overheated and claustrophobic driving cabs made them unpopular with crews and they all had been withdrawn by 1948.

Push-pull trains

While the GWR's steam railmotors proved successful they had severe limitations – their lack of power proved problematical on steeply graded lines and they were virtually incapable of hauling a loaded trailer. To overcome this the GWR equipped some of its more powerful tank engines with auto-control gear that could be operated by the driver remotely from a driving compartment at the front of a trailer coach – trains could sometimes be made up of four coaches with the locomotive in the centre. First introduced in 1904 these trains operated in a push-pull mode and by the 1920s could be seen in operation on GWR routes in the West Country, Gloucestershire, South Wales and out of Paddington. The GWR's most successful auto-fitted tank locomotives were the '1400' Class 0-4-2

that were introduced in 1932 and which, with their 5-ft 2-in. driving wheels, were capable of speeds in excess of 70 mph. The last examples were withdrawn in 1965.

Other railway companies soon followed the GWR and by 1906 the North Eastern Railway, the London, Brighton & South Coast Railway, the Midland Railway and the Great Central Railway were all operating push-pull passenger trains. Push-pull trains remained an important part of railway life on all the 'Big Four' company networks through to the British Railways era in the 1950s when they were gradually replaced by new diesel multiple units.

Diesel railcars

Apart from the Great Western Railway the other three of the 'Big Four' railway companies experimented with diesel-electric railcars in the inter-war years. The LNER, LMS and Southern Railway all took delivery of a 250-hp railcar each built by Armstrong Whitworth while in 1938 the LMS built a 3-car 250-hp articulated diesel multiple unit which ran successfully on the Varsity Line between Cambridge and Oxford until 1940, the war effectively ending any further development.

It was the GWR that took the bold step of introducing diesel railcars on a wider scale. Powered by AEC engines, thirty-eight of various variants were introduced in total between 1933 and 1942 – some of these were twin-sets fitted with a buffet and toilets that were highly successful in capturing passenger traffic on the Birmingham (Snow Hill) to Cardiff route. Extremely cost-effective single railcars were also successfully used on rural branch lines and became very popular with passengers for their comfort, cleanliness and smooth running. Remaining in service until the early 1960s, their eye-catching streamlined shape and brown and cream livery gave them the nickname of 'Flying Bananas'.

LEFT: The LNER's steam railcar No.92 *Chevy Chase* seen new at Doncaster Works in 1928. Eleven of these were built by Clayton Wagons Ltd of Lincoln but had a short working life and had been withdrawn by 1937.

BELOW: Built by the Gloucester Railway Carriage & Wagon Company in 1936, ex-GWR streamlined diesel railcar W8 is seen here at Welshpool in 1953 while on a Railway Development Society trip.

Locomotive works

In 1923 the 'Big Four' inherited a motley collection of locomotives and rolling stock from their predecessors, including a varied number of locomotive, carriage and wagon (C&W) works. During the Second World War production was switched to the war effort and while their menfolk were fighting overseas women took on much of the work, manufacturing parts for Merlin engines, tanks, gliders and armaments. During this period locomotives and rolling stock suffered greatly from lack of maintenance, the immediate post-war years seeing Britain's railway network in poor shape.

Great Western Railway (GWR)

The GWR's main works had been established at Swindon in 1841 and under successive locomotive superintendents – Daniel Gooch, Joseph Armstrong, William Dean, George Jackson Churchward and Charles Collett – had turned out many highly successful passenger and freight locomotives. The nineteenth century broad-gauge single-wheeler 2-2-2s and 4-2-2s were followed by standard gauge 'City' and 'County' 4-4-0s and 'Star' and 'Saint' 4-6-0s of the early twentieth century. The 1920–30 period was the golden age at Swindon when Charles Collett's 'King', 'Castle' and 'Hall' 4-6-0s rolled off the production line, the workforce standing at 14,000. Following nationalization Swindon continued building steam locomotives to GWR designs and also 198 of the BR standard types until 1960. Construction of diesel hydraulic locomotives and diesel multiple units continued until 1965 but the famous works finally closed in 1986.

Elsewhere on the GWR system the Wolverhampton Works had been somewhat eclipsed by the Swindon post-broad-gauge era in 1892. After producing 800 locomotives, the works ceased building in 1908 but continued to repair and overhaul until 1964. The former Rhymney Railway's 1899 works at Caerphilly was taken over by the GWR and continued to repair and overhaul locomotives until 1963. The former Cambrian Railways' 1865 Oswestry works continued to repair and overhaul steam locomotives until 1966.

London Midland & Scottish Railway (LNER)

The LMS inherited several large works, the largest, at Crewe, being established in 1843 by the Grand Junction Railway. Under successive Chief Mechanical Engineers (CMEs) – John Ramsbottom, Francis Webb, George Whale,

Charles Bowen-Cooke, H.P.M. Beames and George Hughes – it went on to become the main locomotive works for the London & North Western Railway and by 1923 had out-shopped over 6,000 new and rebuilt locomotives. In 1932 William Stanier was appointed CME of the LMS and Crewe went on to build some of the most successful steam locomotive types in Britain, including 'Black 5' and 'Jubilee' 4-6-0s, 'Princess Royal' and 'Coronation' 4-6-2s and '8F' 2-8-0s. Following nationalization Crewe built many standard classes including 'Britannia' and 'Clan' 4-6-2s and '9F' 2-10-0s. Diesel-and electric locomotive building continued until 1990 by which time Crewe had built over 8,000 locomotives, employing 20,000 workers at its peak.

The Midland Railway's 1844 Derby locomotive works also came under LMS management in 1923. While the majority of the LMS's larger engines were built at Crewe, Derby Works built many tank locomotives and during the war a large batch of 'Black 5' 4-6-0s. Steam locomotive building continued until 1957 when the 2,941st locomotive rolled off the production line. Large numbers of diesel locomotives were also built between 1958 and 1962 while the adjacent C&W Works at Litchurch Lane made some of the first diesel multiple units in Britain. Although the main works closed years ago the latter continues to make rolling stock under private ownership.

Several other locomotive works were inherited by the LMS. The former Caledonian Railway's 1856 works at St Rollox, Glasgow, continued to repair and overhaul locomotives under LMS and BR management until 1986. In Lancashire the former Lancashire & Yorkshire Railway's 1886 Horwich locomotive works continued to build steam locomotives until 1957 under LMS and BR management. Horwich works closed in 1983.

LEFT: 'King' Class 4-6-0 No. 6001 *King Edward VII* at speed on rollers at the GWR's Locomotive Testing Plant in Swindon Works in the 1930s. For years this was the only plant of its kind in Britain where the power output of steam locomotives could be assessed accurately.

BELOW: Designed by George Hughes for the Lancashire & Yorkshire Railway, two '5P' Class 4-6-4 tank locomotives are seen here in Horwich Works erecting shop on 10 August 1926.

London & North Eastern Railway (LNER)

The LNER inherited locomotive works at Doncaster, Stratford, Cowlairs, Darlington, Gateshead, Gorton and Inverurie. The Great Northern Railway 1853 Doncaster Works went on to build some of the finest and most graceful steam locomotives in the world under the LNER's CME Nigel Gresley, during the 1930s turning out 'A3' and 'A4' 4-6-2s and the highly successful 'V2' 2-6-2s. Following Gresley's death in 1941 the works built Thompson 'A2/2's and Peppercorn 'A1' Pacifics, many of the latter built by BR. The post-war years saw production of BR standard locomotives, the last being out-shopped in 1957 by which time Doncaster had built over 2,000 steam locomotives. Diesel-electric and electric locomotive building continued until 1987.

At Stratford in East London, the Eastern Counties Railway had opened the Great Eastern Railway's locomotive works in 1847. Locomotive building ceased soon after the LNER took over, by which time Stratford had built 1,700 steam locomotives. Under LNER and BR management the works repaired and overhauled steam and diesel locomotives until 1963. A new diesel repair shop survived until 1991. Much of the site was used for the London 2012 Olympic Games.

Opened by the Edinburgh & Glasgow Railway in 1841, the former North British Railway's locomotive works at Cowlairs in Glasgow ceased locomotive building following LNER takeover. It continued to repair and overhaul steam and later diesel locomotives until 1994. The former 1903 Great North of Scotland Railway's locomotive works at Inverurie in Aberdeenshire had already ceased production in 1921 but repaired and overhauled locomotives for both the LNER and BR before closure in 1969.

In northeast England the LNER inherited the former North Eastern Railway's Stockton & Darlington Railway 1863 works at Darlington. The LNER continued building steam locomotives including Gresley's 'hush-hush' Class 'W1' 4-6-4 in 1929. Following nationalization in 1948 Darlington built Peppercorn 'A1' 4-6-2s and several BR standard loco types and from 1953 until closure in 1966 built large numbers of diesel shunters for BR. At Gateshead the NER's works had ceased production of new locomotives in 1910 but continued to repair and overhaul until closure in 1932.

In Lancashire the former Great Central Railway's Gorton locomotive works had been opened in 1848 by its predecessor, the Manchester, Sheffield & Lincolnshire Railway. By 1923 Gorton had built over 900 steam

locomotives but this slowed to a trickle under LNER and BR management although it supplied sixty-four electric locomotives for the Woodhead Line between 1950 and 1954. Repair and overhaul of steam locomotives continued until closure in 1963.

Southern Railway (SR)

The Southern Railway inherited locomotive works at Ashford, Brighton and Eastleigh. Ashford Works in Kent was opened in 1847 by the South Eastern Railway, the predecessor to the South Eastern & Chatham Railway, which continued locomotive building there until 1923. BR continued to build, repair and overhaul locomotives until closure in 1963 – during the Second World War it produced 20 of Bulleid's utilitarian 'Q1' Class 0-6-0s and a number of Stanier's '8F' 2-8-0s.

On the south coast at Brighton the London & Brighton Railway's works had been established in 1840, later becoming the main works facility for the London, Brighton & South Coast Railway and continuing to build locomotives under SR and BR management – when production ceased in 1957 it had built more than 1,200 steam locomotives including 100 of Bulleid's 'Light Pacifics'.

The London & South Western Railway (LSWR) opened the SR's main locomotive works at Eastleigh in Hampshire in 1910 after a move from Nine Elms in London. Under successive CMEs – the LSWR's Dugald Drummond and Robert Urie, and the SR's Richard Maunsell and Oliver Bulleid – Eastleigh turned out many outstanding locomotive classes including the 'Schools' Class 4-4-0s, the 'King Arthur' and 'Lord Nelson' 4-6-0s and the innovative 'Merchant Navy' 4-6-2s. Following nationalization in 1948 it turned its hand to rebuilding over ninety of Bulleid's air-smoothed 'Pacifics', building third-rail electric multiple units and repairing and overhauling steam locomotives until 1967.

LEFT: A crowd of workmen watch the lifting of a locomotive at the North Eastern Railway's locomotive works in Gateshead in 1910. The works continued to repair and overhaul locomotives until 1932.

ABOVE: Designed by Oliver Bulleid, 'West Country' Class 4-6-2 No. 34007 *Wadebridge* being lifted from its wheels during repairs at Eastleigh Works in October 1949.

Expansion of Southern third-rail electric

On its formation in 1923 the Southern Railway (SR) inherited just eighty-three route miles of electrified railway. This consisted of the former London, Brighton & South Coast Railway's (LB&SCR) South London 'Elevated Electric' AC system that supplied power from overhead wires, and the London & South Western Railway's (LSWR) third-rail DC inner suburban service out of Waterloo. Also absorbed by the SR was the short Waterloo & City Railway, an underground line in London that was completely separate from the rest of the capital's underground system.

From the start the SR management favoured the third-rail DC system and by 1925 this had spread out from Waterloo to Guildford, Dorking and Effingham, and from Holborn Viaduct and Victoria to Orpington via the Catford Loop. The following year third-rail electrification had been extended along the South Eastern Main Line to Orpington and the three routes to Dartford. In the same year the SR announced officially that the existing AC overhead system was going to be scrapped in favour of third rail and the last of the former LB&SCR's 'Elevated Electric' trains ran in 1929, by which time the SR was operating just under 300 route miles of third-rail electrified track.

The 1930s was a period of expansion for the SR's electrification plans. By the outbreak of the Second World War, the former LSWR main lines from Waterloo to Portsmouth and Reading had been electrified, as had the LB&SCR's mainline from Victoria to Brighton, Hastings and Eastbourne, and the South Eastern & Chatham Railway's mainline to Sevenoaks and Maidstone. The war brought a temporary halt to further electrification but this resumed once the dust had settled, with the Kent Coast routes being electrified by 1961. At the same time the voltage across the whole of the BR (Southern Region) network was increased from 600V DC to 750V DC. The South Western Main Line from Waterloo to Southampton and Bournemouth was electrified in 1967 and extended to Weymouth in 1988. Other routes electrified during the 1980s were the Hastings Line, East Grinstead branch and Eastleigh to Fareham.

BELOW: Southern Railway electric multiple units, such as this 2-BIL unit No. 2116 which was built at Eastleigh Works, entered traffic in the 1930s on the newly electrified Brighton, Portsmouth and Medway routes.

'Brighton Belle'

The London, Brighton & South Coast Railway introduced Pullman cars in 1875. The company introduced the first all-Pullman train in the UK some six years later, the 'Pullman Limited' operating services between London Victoria and Brighton. The LB&SCR introduced a new Pullman train between Victoria and Brighton in 1908. Named the 'Southern Belle', this train made two return journeys each way on weekdays, completing the 50½-mile journey in exactly sixty minutes. It continued to operate as a steam-hauled service up until 1933 upon third-rail electrification of the London to Brighton. Three five-car all-Pullman electric multiple units were then introduced and in 1934 the service was renamed the 'Brighton Belle'. The service was suspended during the Second World War but reinstated in 1946. Its age was beginning to show by the early 1960s when the train was operating three return journeys each weekday and two on Sundays. Despite excellent patronage, the ageing electric units were nearing the end of their life and British Railways withdrew the service on 30 April 1972. Progress is now being made in restoring two of the driving cars and reuniting them with their preserved coaches so that the 'Brighton Belle' can run once again on special charter services.

BELOW: The all-electric 'Brighton Belle' was one of four Pullman trains operated by the Southern Region of British Railways in the late 1940s and early 1950s. It was also the most long lived, being withdrawn from service in 1972. A complete 5-car set is currently being restored for main-line running in the very near future.

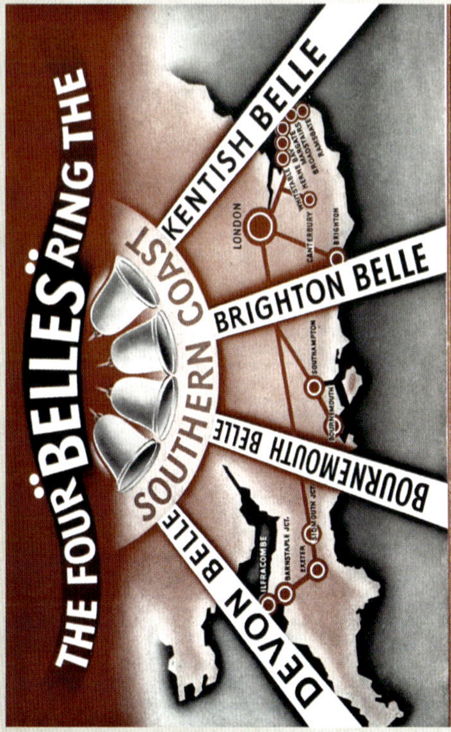

Minor cross-country railways

Midland & South Western Junction Railway (M&SWJR)

One of the crazy railway schemes of the 1840s 'Railway Mania' period was the grandly named Manchester & Southampton Railway, which had been put before Parliament by the London & Birmingham Railway. Seen as an important north-south cross-country route that left the Midland Railway's mainline at Cheltenham and joined the London & South Western Railway (LSWR) at Romsey, like many railway schemes of this period, it was rejected by Parliament in 1848.

Its successor was not such a grand affair. Nicknamed the 'Tiddley Dyke', the single-track Midland & South Western Junction Railway between Cheltenham and Andover was formed in 1884 by the amalgamation of two existing railway companies – the Swindon & Cheltenham Extension Railway (S&CER) and the Swindon, Marlborough & Andover Railway (SM&AR). The latter had opened between Red Posts Junction at Andover and Rushey Platt at Swindon in 1884 although shortage of capital meant that the section between Grafton and Marlborough was incomplete. The Swindon & Cheltenham Railway had already opened between Rushey Platt and Cirencester in 1883 but cash problems resulted in the northern section from Cirencester to Andoversford not being completed until 1891, by which time the company had amalgamated with the SM&AR to form the M&SWJR. Running powers were obtained from the GWR enabling M&SWJR trains to run between Andoversford and Cheltenham, terminating at the Midland Railway's Lansdown station.

By the 1890s the M&SWJR was almost bankrupt but was saved when Sam Fay of the LSWR became General Manager in 1892. During his tenure he turned the M&SWJR around, completing the missing link between Grafton and Marlborough and obtaining running powers over the LSWR's 'Sprat & Winkle Line' between Andover and Redbridge. Fay moved to the Great Central Railway in 1899 by which time trains could travel from the North and the Midlands to Southampton via the M&SWJR – a slow journey with the weekdays through carriage between Liverpool and Southampton taking an excruciating 7½ hours.

The M&SWJR provided a link between military centres at Tidworth and Ludgershall on the Salisbury Plain with Southampton Docks, and saw heavy use during the South African Wars and the First World War. Also important in the railway's early years were racehorse traffic in the Marlborough area, milk traffic north of Swindon and pigeon trains from the North. The M&SWJR became part of the GWR at the 1923 Grouping. By the 1950s services had become considerably reduced with only one through train each weekday between Cheltenham and Southampton Central, taking around four hours to complete the 94½-mile journey. In 1958 this train was diverted to run into Cheltenham (St James) station, the option for changing trains at Lansdown having been lost. By then the writing was on the wall for this delightful rural line and it closed on 10 September 1961.

Stratford-upon-Avon & Midland Junction Railway (S&MJR)

The Stratford-upon-Avon & Midland Junction Railway was formed in 1910 by the amalgamation of four existing railway companies, the Northampton & Banbury Railway, East & West Junction Railway, Evesham, Redditch & Stratford-upon-Avon Railway and the Stratford-upon-Avon, Towcester & Midland Junction Railway, which had opened at various times between 1866 and 1891. Despite offering an east-west cross-country route between Broom Junction in Warwickshire, on the Midland Railway's route between Evesham and Redditch, and Blisworth, on the West Coast Main Line near Northampton, they were all on the verge of bankruptcy upon amalgamation.

The new railway became an important through route for Midland Railway freight traffic between Bristol and St Pancras via Ashchurch and Broom Junction. Local passenger traffic amounted to a mere three return journeys on weekdays between Blisworth and Stratford and its worn-out, decrepit track soon earned it the nickname of the 'Slow, Miserable and Jolty'.

During the post-1923 LMS period the S&MJR not only grew as an important east-west route for freight traffic bypassing Birmingham but also for excursion traffic. The LMS also experimented with a Karrier Ro-Railer converted single-decker bus in 1932 but this innovation was unsuccessful and was withdrawn after a short time. Passenger services were withdrawn between Broom Junction and Stratford in 1947 and between Blisworth and Stratford-upon-Avon in 1952. Blisworth to Cockley Brake closed completely in 1951 followed by Blisworth to Ravenstone Junction in 1952. Now under the control of BR Western Region, freight continued to use the S&MJR for some years more, including a growing number of iron-ore trains from Northamptonshire to South Wales which ran via a new 1960 connection with the Honeybourne line at Stratford. From that date the section west to Broom Junction closed completely. Ironstone traffic to South Wales had declined by 1965 and the remaining S&MJR route was closed completely on 5 July with the exception of the section from Fenny Compton that continues to serve the MOD establishment near Kineton.

LEFT: Built for the Midland & South Western Junction Railway by Dübs & Co. in 1894, 2-4-0 No. 1336 was later 'Swindonised' by the GWR. Seen here at rest in Reading engine shed on 23 August 1930, the locomotive survived into the British Railways era and was withdrawn in 1954.

ABOVE: Former Midland Railway '3F' 0-6-0 No. 3698 trundles along the Stratford-upon-Avon & Midland Junction Railway at Boundary Farm Bridge, near Kineton in Warwickshire, with the 1.25 p.m. Blisworth to Stratford train on a sunny 8 June 1939.

Famous named trains

Very few trains were named in the nineteenth century with the exceptions being the Anglo-Scottish rivals of the West Coast Main Line's 'The Royal Scot' and the East Coast Main Line's 'Special Scotch Express' which were both introduced in 1862. The inter-wars of the twentieth century saw a large number of long-distance trains in Britain receiving names – the perceived romance of rail travel of the 1920s and 1930s led the 'Big Four's' publicity departments to produce beautifully designed posters aimed at enticing both businessmen and holidaymakers to use their services. Decorative luggage labels, restaurant-car menus and fold-out leaflets describing the 'view from the window' all added to the passengers' experience.

Always at the forefront of publicity, the Great Western Railway had named its premier express between Paddington and Penzance the 'Cornish Riviera Limited' as early as 1904 but it took until the formation of the 'Big Four' railway companies in 1923 for the fashion of naming trains to really take off. 1927 was the highpoint with eighteen named trains being introduced. Unfortunately the outbreak of the Second World War brought an end to these high-speed luxury trains.

London & North Eastern Railway (LNER)

'Norfolk Coast Express' (1907–1939)
London Liverpool Street to Cromer and Sheringham

'Flying Scotsman' (1862–1923 named the 'Special Scotch Express'/1924–present day)
London King's Cross to Edinburgh Waverley

'Scarborough Flyer' (1927–1939/1950–1963)
London King's Cross to Scarborough

'The Aberdonian' (1927–1987)
London King's Cross to Aberdeen

'The Hook Continental' (1927–1939/1945–1987)
London Liverpool Street to Harwich Parkeston Quay

'West Riding Pullman' (1927–1935 then as 'The Yorkshire 'Pullman' 1935–1939/1946–1978)
London King's Cross to Leeds/Bradford

'Queen of Scots' (1928–1939/1948–1964)
London King's Cross to Glasgow Queen Street via Leeds

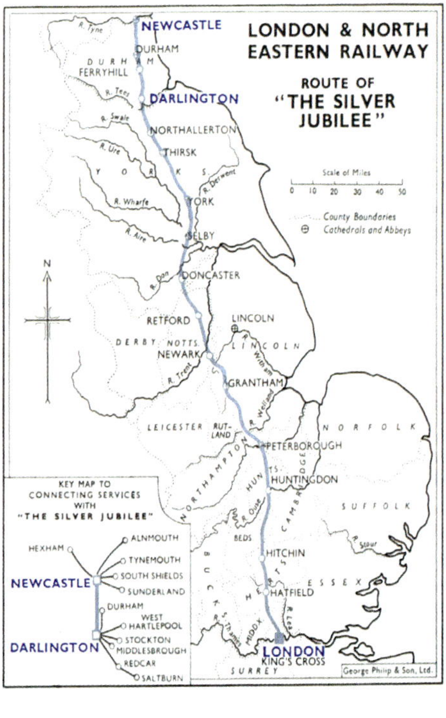

'Cambridge Buffet Expresses' 1932–1939/1945–1978)
London King's Cross to Cambridge
'The Silver Jubilee' (1935–1939)
London King's Cross to Newcastle-upon-Tyne
'The Coronation' (1937–1939)
London King's Cross to Edinburgh Waverley
'The East Anglian' (1937–1939/1946–1962)
London Liverpool Street to Norwich
'West Riding Limited' (1937–1939)
London King's Cross to Leeds/Bradford

London Midland & Scottish Railway (LMS)
'The Royal Scot' (1862–2003)
London Euston to Glasgow Central
'Pines Express' (1927–1939/1949–1967)
Manchester London Road to Bournemouth West
'The Devonian' (1927–1939/1946–1975)
Bradford Forster Square to Paignton and Kingswear
'The Irish Mail' (1927–1985)
London Euston to Holyhead
'The Lakes Express' (1927–1939/1950–1965)
London Euston to Blackpool/Windermere/Workington
'The Mancunian' (1927–1939/1949–1966)
London Euston to Manchester London Road
'The Manxman' (1927–1939/1951–1965)
London Euston to Liverpool Lime Street
'The Merseyside Express' (1927–1939/1946–1966)
London Euston to Liverpool Lime Street
'The Mid-Day Scot' (1927–1966)
London Euston to Glasgow Central
'The Royal Highlander' (1927–1939/1957–1985)
London Euston to Inverness
'The Thames-Clyde Express' (1927–1939/1949–1974)
London St Pancras to Glasgow St Enoch
'The Thames-Forth Express' (1927–1939)
London St Pancras to Edinburgh Waverley
'The Ulster Express' (1927–1939/1949–1975)
London Euston to Heysham
'The Welshman' (1927–1939/1950–1964)
London Euston to Llandudno/Portmadoc/Pwllheli
'The Lancastrian' (1928–1939/1957–1962)
London Euston to Manchester London Road

LEFT: Introduced between London King's Cross and Newcastle in 1935 to commemorate twenty-five years of reign by King George V, the LNER's 'Silver Jubilee' was Britain's first streamlined high-speed express. Hauled by one of Nigel Gresley's new 'A4' locomotives, the train consisted of two pairs of articulated coaches separated by a triplet set – engine and coaches were finished in two-tone silver and grey with stainless steel embellishments.

BELOW LEFT: A luggage label from the LNER's streamlined 'Coronation' train which was introduced in 1937 between King's Cross and Edinburgh.

BELOW: A luggage label from the LNER's 'Yorkshire Pullman' which when introduced in 1935 conveyed coaches from King's Cross to Bradford, Halifax and Harrogate.

'The Comet' (1932–1939/1949–1962)
London Euston to Manchester London Road
'The Grampian Corridor' (1933–1939/1949–1967)
Glasgow Buchanan Street/Queen Street to Aberdeen
'The Granite City' (1933–1939/1949–1968)
Glasgow Buchanan Street/Queen Street to Aberdeen
'Bon Accord' (1937–1939/1949–1968)
Glasgow Buchanan Street/Queen Street to Aberdeen
'Coronation Scot' (1937–1939)
London Euston to Glasgow Central
'The Saint Mungo' (1937–1939/1949–1968)
Glasgow Buchanan Street/Queen Street to Aberdeen
'The Palatine' (1938–1939/1957–1964)
London St Pancras to Manchester Central

Great Western Railway (GWR)
'Cornish Riviera Limited' (1904–1917/1919–1946/1947–present day)
London Paddington to Penzance with numerous slip coaches
'Cheltenham Flyer' (1923–1939)
London Paddington to Cheltenham
'Torbay Express' (1923–1968)
London Paddington to Paignton and Kingswear
'Cambrian Coast Express' (1927–1939/1951–1967)
London Paddington to Aberystwyth/Pwllheli
'The Bristolian' (1935–1939/1954–1965)
London Paddington to Bristol

Southern Railway (SR)
'Southern Belle' (1908–1933)
London Victoria to Brighton
'The Thanet Belle' (1921–1939/1948–1950)
London Victoria to Ramsgate Harbour
'Atlantic Coast Express' (1927–1939/1945–1964)
London Waterloo to Torrington/Ilfracombe/Bude/Padstow/Plymouth
'Golden Arrow' (1929–1939/1946–1972)
London Victoria to Paris Nord
'The Bournemouth Belle' (1931–1939/1946–1967)
London Waterloo to Bournemouth West
'Brighton Belle' (1934–1939/1946–1972)
London Victoria to Brighton
'The Night Ferry' (1936–1939/1947–1980)
London Victoria to Paris Nord

BELOW: Rebuilt 'Battle of Britain' Class 4-6-2 No. 34088 *213 Squadron* heads the 'Golden Arrow' Pullman train through Ashford on 13 August 1960. The train was introduced by the Southern Railway in 1929 and, apart from the years of the Second World War, continued to operate until 1972.

Pullman trains

Luxury Pullman sleeping cars were first introduced in the USA by George Mortimer Pullman in 1864. His Pullman Company went on to become a great success and the idea was exported to Britain, where the Midland Railway introduced Pullman cars in 1874. Soon other British railway companies were following suit and in 1882 the Pullman Car Company of Britain was established at Brighton. The first company to operate a complete Pullman train was the London, Brighton & South Coast Railway, which introduced the *Pullman Limited* between London Victoria and Brighton in 1881. The familiar umber and cream livery began to be adopted for Pullman coaches in 1906. Probably the most famous of British Pullman trains were the Southern Railway's *Bournemouth Belle*, introduced in 1931, and the all-electric *Brighton Belle*, introduced in 1934 – the former continued to run during the BR-era until 1967 and the latter until 1972.

The LNER introduced the 'West Riding Pullman' in 1927, this train being renamed 'The Yorkshire Pullman' in 1935. The company also introduced 'The Queen of Scots' Pullman train between King's Cross and Glasgow Queen Street in 1928. Discontinued during the Second World War, both trains were reintroduced after the war with the former being withdrawn in 1978 and the latter in 1964.

BELOW: Produced for the Southern Region of BR in 1948, this attractive poster was painted by Charles 'Shep' Shepherd and shows the four named Pullman trains then in operation across the region. Introduced in 1947, the summer-only 'Devon Belle' ceased in September 1954.

Railway publicity

Britain's railway companies had discovered the value of marketing their services, routes and destinations to the travelling public by the beginning of the twentieth century. Railway publicity departments all over the land were extolling their companies' virtues with illustrated route maps, guidebooks, postcards and posters long before the First World War. The publicity departments of the North British Railway and the London & North Western Railway were noteworthy, but smaller companies such as the North Staffordshire Railway and the Furness Railway also led the way. The Great Northern Railway's 'Skegness is So Bracing' poster also placed it at the forefront of publicity – designed by the illustrator John Hassall in 1908, it is probably the most enigmatic railway poster of all time.

However, the Great Western Railway (GWR) had by far the most prolific publicity department which, as early as 1903, was promoting its route to the southwest as 'The Holiday Line'. The GWR even commissioned a film, 'The Story of the Holiday Line', which was first screened at the London Coliseum in 1914, and was promoting the scenic delights of Devon and Cornwall as the 'Cornish Riviera' by the 1920s – a marketing slogan it used to good effect on posters, guide books and even on its premier express to Penzance. The GWR's 'Cheltenham Flyer' – dubbed the 'World's Fastest Train' in the early 1930s – was a tremendous coup for its publicity department, as was the centenary of the company in 1935. Adorning stations the length and breadth of GWR's system could be seen posters by artists such as Murray Secretan, Leonard Richmond and Charles Mayo. Not content with its prolific output of holiday guides and posters the GWR also produced over forty jigsaw puzzles (manufactured by Chad Valley) and a series of books for railway enthusiasts.

The Golden Age of Railways in Britain was in the 1930s and the publicity departments of the other 'Big Four' railway companies – the London, Midland & Scottish, the London & North Eastern and the Southern railways – were also engaged in glamourizing travel on their respective systems. Undoubtedly the most prolific poster artist of this period was Norman Wilkinson, who produced a large number of large-format classic posters for both the LMS and the SR. In addition to its posters featuring holiday destinations, the Southern Railway also produced a series of attractive railway walks books written by the author and broadcaster S.P.B. (Stuart Petre Brodie) Mais. Always ready to capitalize on a publicity coup, the LNER's department had a field day when that company introduced Britain's first streamlined high-speed train in 1935, the 'Silver Jubilee', between London King's Cross and Newcastle. Stylish locomotives and matching coaches ushered in a new age of modern travel, complemented by stylish corporate posters and brochures. The LNER's new streamliner, 'The Coronation', was by 1937 competing with the LMS' new streamlined 'The Coronation Scot' for the important Anglo-Scottish traffic, with both companies vying with each other for the ultimate publicity coup – the LNER was the winner when in 1938 one of their streamlined locomotives, *Mallard*, achieved a world record for steam traction. The Second World War broke out just over a year later, and a new age of austerity descended upon the railways.

BELOW: John Hassall's famous poster. 'Skegness is So Bracing' was produced for the Great Northern Railway in 1908. Hassall was apparently paid 12 guineas for the artwork – the original now hangs in Skegness Town Hall.

Norman Wilkinson

Born in Cambridge in 1878, Norman Wilkinson, after receiving an early musical training, went on to study at Southsea School of Art and became an accomplished maritime artist, illustrator, poster artist and inventor of wartime camouflage. His illustration career started in 1898 when he was commissioned by the *Illustrated London News* and he went on to design shipping posters for companies such as Cunard and a vast number of railway posters, firstly for the London & North Western Railway and then its successor, the LMS. He hit upon the idea of camouflaging naval ships when serving with the Royal Naval Volunteer Reserve during the First World War. His ideas were so successful that he led a team of camouflage artists in both World Wars and received an award for his invention of 'dazzle painting'.

As well as becoming one of the finest maritime painters of the twentieth century (his painting, 'Plymouth Harbour', hung in the First Class smoking room of the *Titanic*), Wilkinson is famous for his large-format landscape railway posters that adorned stations up and down the country. He even persuaded seventeen Royal Academicians to produce similar posters for the LMS in 1924. Wilkinson died in 1971 after a long and illustrious career as one of Britain's most successful twentieth-century commercial artists.

Railways at war (2)

As we have seen in the previous chapter, the Railway Executive Committee (REC) took over control of Britain's railways during the First World War. With the clouds of war again looming over Europe, the REC was re-formed on 24 September 1938 and less than a year later, just two days before Britain declared war on Germany, the REC again took control of Britain's railways. The Government then made fixed annual payments in exchange for the net revenues of the railway operating companies. REC control lasted through the war and until 1 January 1948 when the railways were nationalized.

Despite blackouts, major disruption caused by bombing, lack of maintenance and overworked staff and locomotives, Britain's railways kept running through the war. During the last four months of 1939 the railways not only operated 4,349 special trains carrying nearly 1.5 million children and civilians away from London during the first evacuation of the capital, but also carried 390,000 troops of the British Expeditionary Force (BEF) in 1,100 special trains to Channel ports for embarkation to France. May and June 1940 saw the BEF rescued from Dunkirk and in a period of a fortnight the railways ran 620 special trains to carry the 319,116 troops away from the Channel ports.

The railways suffered heavily during the German bombing of London, other cities and key installations – stations, junctions, goods yards and engine sheds were all targeted but despite suffering considerable damage the railways kept running. During the German bombing of Coventry on 14 November 1940, when 400 Luftwaffe aircraft targeted the city, no fewer than 122 of the 600 incidents reported were on railway property. Coventry station, sidings, junctions, main and branch lines, bridges and viaducts were all hit but within two days both the Coventry to Birmingham and Coventry to Leamington lines had been reopened.

Troop trains, workmen's trains and freight trains all had priority over passenger trains, where delays and overcrowding were normal. Wartime restrictions saw passenger trains reduced in number, speeds restricted, seat reservations cancelled, sleeping car trains considerably reduced and restaurant cars withdrawn. By 1944 nearly 7,000 special trains were run every day to carry workers to Government factories and in one year alone 385 million passenger journeys were made by holders of workmen's tickets. Of course the war effort not only depended on the mining and rail distribution of coal, which by 1942 amounted to 160 million tons a year, but also the distribution of other raw materials, war equipment and munitions between war factories and rail-connected docks.

Women played an important part in keeping Britain's railways moving. They took over many jobs normally

carried out by men including maintenance of locomotives and rolling stock, working on the permanent way, making concrete sleepers, operating cranes, and working as signalwomen, guards, motor van drivers, woodworkers and metalworkers.

William Stanier's '8F' 2-8-0 was chosen by the War Department as the standard heavy-freight locomotive during the war. A total of 852 were built with many seeing service in war theatres overseas especially in the Middle East and Egypt. These were supplemented by 935 'Austerity' 2-8-0 locomotives designed by Robert Riddles for the War Department and built by the North British Locomotive Company in Glasgow and Vulcan Foundry in Lancashire. Introduced in 1943, many saw service on Europe's war-damaged railways following the Allied invasion of 1944.

The lead-up to the D-Day landings in 1944 saw US Army Transportation Corps locomotives shipped in to Britain to assist with the largest mass movement of troops and military equipment in the history of warfare. By 3 June the railways had moved in great secrecy 500,000 British troops, 2 million US troops and 1.5 million tons of stores from their concentration areas to ports of embarkation along the south coast.

The Second World War left a terrible legacy for Britain's railways, their rundown state and lack of investment leading inevitably to nationalization in 1948.

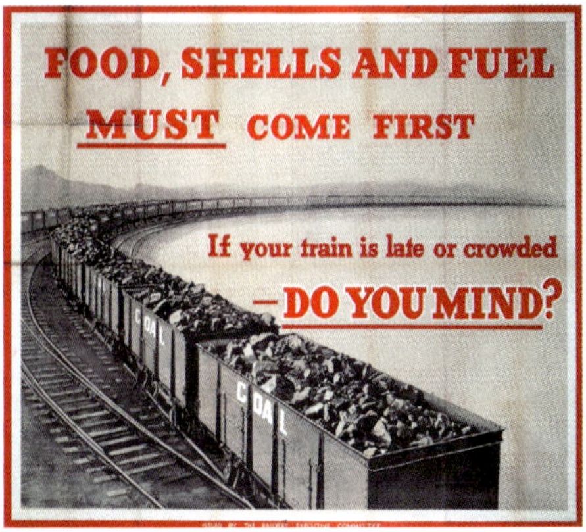

LEFT: Repairing the damage – an LNER 'B12' 4-6-0 slowly hauls its long train over a recently repaired bridge between Liverpool Street and Stratford that had been damaged by a German flying bomb in 1944.

TOP: A trainload of US Sherman tanks waits in the Great Western Railway's Acton Railway Yard No. 2 before being hauled to an unknown destination. In the lead up to D-Day Britain's railways moved in great secrecy over 2.5 million troops and 1.5 million tons of equipment to south coast ports.

ABOVE: Poster produced by the Railway Executive Committee during the Second World War to highlight wartime priorities. Moving food, munitions and fuel took priority over passengers.

Railway women

It is now recognized that women had a very tough time in the workplace until the latter quarter of the twentieth century. They received much lower remuneration than their fellow men, working conditions were often abysmal, trade-union recognition was very slow in coming and job perks never came their way. It was a man's world and especially so on Britain's railways. Women working on the railways were looked upon as cheap labour and usually had the most menial jobs such as office clerks, waiting room attendants, laundry workers, workshop seamstresses or catering assistants. There were a handful of instances of station mistresses at small country stations but, despite this relatively responsible position, women were still paid a fraction of their male counterparts.

By the outbreak of the First World War Britain's railways employed around 630,000 people, of whom only 13,000 were women, all working in low-pay jobs – important posts such as engine drivers, firemen, locomotive cleaners, signalmen, locomotive maintenance and repair were all 100 per cent male dominated. One of the most responsible jobs that women held on the railways at that time was as a level-crossing keeper – the women were usually wives or widows of railwaymen and the post brought with it the tenancy of a railway-owned cottage, enabling them to carry on with their domestic work in between the closing and opening of gates. By the outbreak of war just under half of all level crossings on the Great Western Railway network were operated by women. The job was badly paid with long hours but it brought with it an assured tenancy in a crossing-keeper's cottage.

Recognition of trades unions by the railway companies was slow in coming and union recognition of female employees even slower. The Associated Society of Locomotive Engineers & Fireman (ASLEF) and the United Pointsmen's & Signalmen's Society (UPSS) refused to admit women because they were already barred from holding any of these positions by the railway companies. The Amalgamated Society of Railway Servants (ASRS) also barred women from joining. Up until the First World War only the Railway Clerks' Association (RCA) and the General Railway Workers Union (GRWU) allowed women to join and only if they were wages grade. In 1913 the GRWU, ASRS and UPSS combined to become the National Union of Railwaymen (NUR) but despite its rules allowing females to join it still barred their membership.

The First World War saw a major, but temporary, shift in attitude to female railway workers. With their menfolk joining up to fight abroad women were allowed to replace them in some uniformed posts but were expected to stand down once the war was over. Recruitment started in

1915 and soon women had taken over what were once male-dominated posts such as engine and carriage cleaners, passenger and goods porters, ticket collectors, horse-van drivers and coal shovellers, while there was a tenfold increase in the number of women employed in railway offices. As the war progressed women were also employed as signalwomen and guards. Sadly, women never received equal pay and their treatment by the largest rail unions, the NUR and ASLEF, was particularly shabby. By the end of the war there were nearly 66,000 railway women, a fivefold increase since 1914.

The end of the war saw a return to pre-war conditions for women on the railways. Many were dismissed to make way for the returning men and those that remained were given the most menial of jobs. During the next twenty years the trade unions did little to help, equal pay was never achieved and women in male-dominated positions gradually ebbed away. Low-paid female crossing keepers still closed and opened level crossing gates and in more remote parts of Britain a few station mistresses could still be found, but they were the exception.

The Second World War saw a repeat of what had happened a quarter of a century before. Women joined the railways in their thousands and by 1943 there were nearly 89,000, the vast majority of them in previously male-dominated roles as signalwomen, train guards, workshop women, fitters, maintenance workers, mechanics, porters and even tracklayers. Despite this, although some women became locomotive cleaners they were never allowed to drive or fire one. Although reasonably well paid compared to other civilian jobs, women never reached pay parity with men during the war.

Once the war had ended the new Labour government ignored demands for equal pay for railway women, and women in general. The immediate post-war years saw a repeat of the end of the previous war, when railway women in male-grade posts were summarily dismissed to make way for returning men. By 1947 the number of women employed by the four railway companies had nearly been halved but a manpower shortage soon saw many of them offered their old jobs back.

The three decades following the 1948 nationalization of the railways brought a massive reduction in the number of people, both men and women, employed by British Railways. Thousands of miles of railway and thousands of stations were closed but some women still managed to retain jobs in carriage cleaning, catering, signalling, engineering and, as always, as crossing keepers – all posts still paid significantly less than their male counterparts but with many examples of decades of loyalty to the railways.

Two major pieces of legislation eventually forced the railways to accept women on equal terms: the Equal Pay Act 1970 and the Sex Discrimination Act 1975 were both major steps forward, although the former only came into law at the end of 1975. For the first time in 150 years railway women were treated as equals and within a few years they were in previously male-only posts such as ticket collectors, station managers, senior signalwomen and track maintenance crew. In 1978 Britain's first female train driver began work on the London Underground but it was another five years before Anne Winter became the first train driver on British Rail. Since then the number of female train drivers has grown considerably and it is now not uncommon to see one at the controls of a Virgin Pendolino or Eurostar train. Sadly, the unsung heroes of Britain's railway history, female crossing keepers, are now becoming extinct thanks to their replacement by automated level crossings.

FAR LEFT: A group of female carriage cleaners pose with their inspector in front of one of the London & South Western Railway's third-rail electric multiple units at Wimbledon depot in 1917.

LEFT: By 1943 there were nearly 89,000 women working on Britain's railways. Here one of the first women railway guards is seen on duty at Victoria station in London on 19 July of that year.

Charles Collett

Charles Benjamin Collett was born in London in 1871 and after studying at London University was apprenticed at a well-known firm of marine engine builders. In 1893 he joined the Great Western Railway at Swindon where he trained to become a draughtsman and slowly worked his way up the corporate ladder, becoming Manager of Swindon Works in 1912 and Deputy Chief Mechanical Engineer under George Jackson Churchward in 1919. Under Churchward locomotive design at Swindon had made great strides with standardization of components and proven ideas borrowed from French and American engineers that culminated in his elegant four-cylinder 'Star' Class 4-6-0s. Churchward retired in 1922 and Collett became Chief Mechanical Engineer at Swindon. During his tenure at Swindon, Collett further improved Churchward's standard designs with his first, the four-cylinder 'Castle' Class 4-6-0, being a development of the 'Star' Class but with a larger boiler, an increase in cylinder diameter and an increased grate area. First introduced in 1923, a total of 171 were built with the last emerging from Swindon Works as late as 1950. They were a great success, especially with later modifications, and were regular performers on GWR and later BR (Western Region) expresses until the early 1960s.

Collett's 'King' Class 4-6-0s were introduced in 1927. An enlargement of the 'Castle' Class, they were the largest and most powerful express locomotives built by the GWR and were the most powerful 4-6-0s to be built in Britain. Usually seen at the head of the GWR's heavy premier expresses such as the 'Cornish Riviera', they remained in service until 1962. Collett's other great success was the 'Hall' Class 4-6-0, which was a development of Churchward's 'Saint Class'. 259 were built between 1928 and 1943 and remained as the mixed-traffic workhorse on the GWR and BR (Western Region) until 1965. Collett retired in 1941 and died in 1952.

BELOW LEFT: Charles Collett was Chief Mechanical Engineer at Swindon from 1922 to 1941.

BELOW: Collett's most successful locomotive design was the 4-cylinder 'Castle' Class locomotives of which No. 4073 *Caerphilly Castle* was the first when introduced in 1923. The locomotive is appropriately on display at the Swindon Steam Railway Museum.

Sir William Stanier

William Arthur Stanier was born in Swindon in 1876 and after completing his education joined the GWR in 1892 as an apprentice draughtsman at Swindon Works. Like Collett he worked his way up the company ladder, finally becoming Charles Collett's principal assistant in 1922 – his time spent working under Churchward and Collett at Swindon influenced him greatly when he later joined the LMS.

In 1932 the Chief Mechanical Engineer of the LMS, Sir Henry Fowler, retired and Stanier was appointed in his place. With his years of experience at Swindon under his belt he set about reorganizing and standardizing the company's mixed bag of locomotive designs. During his twelve years as CME at Crewe, Stanier introduced a series of highly successful locomotive types including 841 'Black 5' mixed-traffic 4-6-0s, 190 'Jubilee' Class 4-6-0s and 776 '8F' 2-8-0s – the latter remained the backbone of heavy freight work on the LMS throughout the Second World War and into the BR-era. Stanier's most famous locomotive designs were the powerful 'Princess Royal' and 'Coronation' 4-6-2s, which were the mainstay of express passenger work on the West Coast Main Line for nearly thirty years. Stanier was a scientific adviser to the Ministry of Production during the Second World War and was knighted in 1943. He retired a year later and died in 1965.

BELOW: William Stanier's first 'Pacific'-type locomotive was the 'Princess Royal' Class introduced in 1932. Here, No. 46201 *Princess Elizabeth* was captured by the camera of Eric Treacy in 1960 while hauling down 'The Royal Scot' at Harthope near Beattock. This fine locomotive has since been preserved.

Sir Nigel Gresley

Herbert Nigel Gresley was born in Edinburgh in 1876 and moved with his parents at a young age to Derbyshire. He was educated at Marlborough College and apprenticed at the London & North Western Railway's locomotive works at Crewe. He then moved to Horwich, where he trained at the Lancashire & Yorkshire Railway's locomotive works, before becoming Superintendent of the Carriage & Wagon Department of the Great Northern Railway in 1905. Continuing up the corporate ladder, Gresley succeeded H.A. Ivatt as Locomotive Engineer of the Great Northern Railway (GNR) in 1911. When the GNR became part of the newly formed LNER in 1923, he was appointed Chief Mechanical Engineer at Doncaster Works.

During his time at Doncaster Gresley designed twenty-six locomotive types for both the GNR and LNER. He was responsible for designing the former's 'A1' Class 4-6-2 in 1922 and the LNER's 'A3' Class 4-6-2, introduced in 1927, and the 'V2' Class 2-6-2, introduced in 1936. By far his most famous design was the streamlined 'A4' Class 4-6-2s which first appeared in 1935. One of these, No. 4468 *Mallard*, still holds the 126-mph world speed record for steam traction, which was set in 1938.

BELOW LEFT: Photographed just before his untimely death in 1941, Sir Nigel Gresley was probably the foremost locomotive designer in the world during the 1930s.

BELOW: Gresley's 'P2' Class 2-8-2 locomotive No. 2001 *Cock O' the North* is seen here fresh out of Doncaster Works on 11 July 1934. Built to haul heavy passenger trains on the steeply graded route between Edinburgh and Aberdeen, it was later rebuilt as an LNER Thompson Class A2/2 4-6-2 and withdrawn from service in 1960 before being scrapped. A new-build replica of the 'P2' is currently under construction.

Oliver Bulleid

Oliver Vaughan Snell Bulleid was born in Invercargill, New Zealand, in 1882. Following his father's death, 7-year-old Oliver returned with his mother to Britain and on completion of his education began an apprenticeship with the Great Northern Railway (GNR) at Doncaster Works. In 1905 he was promoted to works manager but three years later went to Paris to work for the Westinghouse Electric Corporation as a test engineer.

In 1910 Bulleid returned to Britain and two years later rejoined the GNR as assistant to the new Chief Mechanical Engineer (CME), Nigel Gresley. Following war work he became Gresley's assistant once again on the formation of the LNER in 1923, a post he held until 1937 when he was appointed as CME of the Southern Railway (SR) following the retirement of Richard Maunsell.

During his tenure as CME of the SR, the innovative Bulleid introduced, under wartime conditions, his ground-breaking, air-smoothed, 'Merchant Navy' Class 4-6-2. Built between 1941 and 1949 using the latest technology these thirty locomotives were followed by forty of the utilitarian Class 'Q1' 0-6-0 mixed traffic locos in 1942 and by 100 of his lightweight 'West Country' and 'Battle of Britain' Class 4-6-2s between 1945 and 1951. Sadly his experimental 'Leader' Class of articulated steam locomotives built between 1946 and 1949 was not a success.

Leaving Britain behind once more, Bulleid was appointed as CME of Córas Iompair Éireann (Irish Railways) in 1949. Here he oversaw the early introduction of diesel-electric locomotives and experimented with an unusual turf-burning steam locomotive. He retired in 1958 and died in 1970.

LEFT: Oliver Bulleid was Chief Mechanical Engineer of the Southern Railway between 1937 and 1948.

BELOW: Built under utilitarian wartime conditions at Eastleigh and introduced in 1941, the third of Bulleid's innovative 'Merchant Navy' Class 4-6-2s was No. 21C3 *Royal Mail*. Rebuilt without its air-smoothed casing in 1959 it was withdrawn in 1967 and has the dubious honour of being the last British steam locomotive to exceed 100 mph when in service.

NATIONALIZATION, RATIONALIZATION and MODERNIZATION

1948–1962

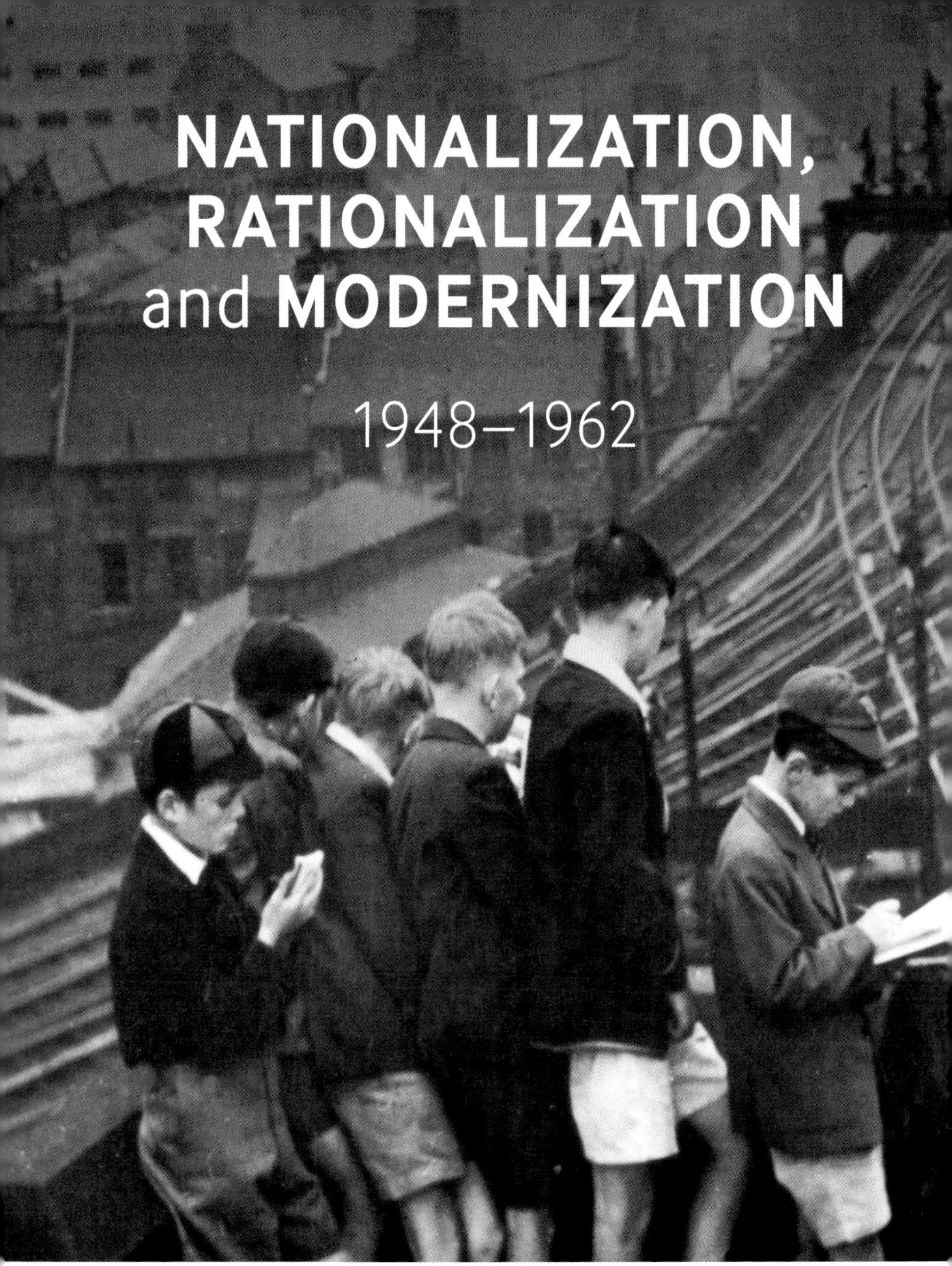

1948	1949	1950	1951	1952	1953	1954	1955

- **1948** – Britain's railways nationalized
- **1948–1949** – Narrow-gauge Corris Railway closed
- **1950** – First gas turbine locomotive enters service
- **1951** – Talyllyn Railway saved from closure by preservationists
- **1952** – 112 people killed in multiple train collision at Harrow & Wealdstone
- **1954** – Manchester to Sheffield via Woodhead route electrified
- **1955** – 'Modernisation Plan' for British Railways published by BTC
- **1955** – National rail strike

1956	1957	1958	1959	1960			1961	1962
'Ferret and dartboard' crest introduced as BR corporate identity	90 people killed in train collision at Lewisham	First diesel-hydraulic locomotives introduced by Western Region	'Condor' express container train introduced	'Blue Pullmans' introduced	Bluebell Railway opens	Last steam locomotive built by British Railways	Type 5 'Deltic' diesel electric locomotives introduced	North British Locomotive Company bankrupt

The birth of British Railways

Government control of Britain's railways during the Second World War had been in the hands of the Railway Executive Committee – this control actually lasted for over eight years from 30 August 1939 until 31 December 1947. The war had left Britain's railways virtually bankrupt and in very poor shape with infrastructure, track, locomotives and rolling stock all suffering from wartime damage and lack of investment and maintenance. While nationalization of the railways had been on the cards following the First World War the problem then was solved by grouping 120 railway companies into just four – the 'Big Four' Grouping of 1923 saw the newly enlarged GWR, LMS, LNER and SR take over a similarly run-down system and by the 1930s Britain's railways were among the best in the world.

However, the concept of nationalization was not rejected following the end of the Second World War. Once the dust had settled on a war-ravaged country, Clement Attlee's Labour Government enacted the Transport Act 1947, which was given royal assent on 6 August that year. Under the Act railways (including their shipping interests), canals, ports, harbours, bus companies and long-distance road haulage were acquired by the state and put under the control of the newly formed British Transport Commission – each form of transport had its own executive committee, with the new British Railways being the commercial name of the Railway Executive. This ambitious plan was admirable, bringing together all forms of land transport to form an integrated system but in reality nothing much changed.

And so on 1 January 1948 British Railways was born. While it was divided into six geographical regions – Southern, Western, London Midland, Eastern, North Eastern and Scottish – it did not include industrial railways, the London Underground, which was taken over by the London Transport Executive, or strangely, the Liverpool Overhead Railway or the narrow-gauge Talyllyn Railway in West Wales, both of which remained independent. While all well intentioned, nationalization did not stop inter-regional rivalry, with the likes of the Western Region more-or-less continuing in the same autonomous fashion from Swindon as its predecessor, the GWR.

The new British Railways (BR) not only inherited a motley collection of thousands of non-standard steam locomotives, some dating back to the 1870s, along with some very ancient wooden rolling stock, but also a dense network of around 20,000 route miles of railway. Many lines competed with each other, a throwback to the pre-Grouping years of the late nineteenth century, while several thousand miles of loss-making and antiquated rural branch lines were a serious drain on BR's finances.

The British Transport Commission dealt with the latter problem by setting up the Branch Lines Committee, which oversaw the closure of over 230 routes totalling around 3,000 route miles between 1948 and 1962 – a national coal shortage during the harsh winter of 1947 had already seen a suspension of services on some of them. On the plus side the British Transport Commission grasped the nettle in 1951 and approved the construction of eleven new classes of standard steam locomotives to replace the existing ageing fleet – using domestically produced coal to fuel them was then still viewed favourably by the Government instead of importing foreign oil to run new diesel locomotives.

Over the coming years there were also changes, not necessarily for the good, to BR's regional boundaries. Old rivalries surfaced when the former Great Central Railway mainline between Sheffield and London Marylebone was transferred from the Eastern Region to the London Midland Region in 1960. From that date the line was living on borrowed time, with express services between the two cities withdrawn and other services placed in the hands of worn-out steam locomotives – closure in 1966 was inevitable. Boundary changes in 1950 and 1958 saw the transfer of much of the former Somerset & Dorset Joint Railway between Bath and Bournemouth to Western Region management, and Swindon gleefully soon put an end to this much-loved railway, which also closed in 1966. A 1963 boundary change saw the same happen to former Southern Region routes west of Exeter when Western Region management got their own back on this late nineteenth-century back-door incursion into the West Country. Despite nationalization nothing had changed! The North Eastern Region was merged with the Eastern Region in 1967.

Despite all the optimism generated in 1948, the following years saw BR's financial position gradually deteriorate and union-organized national rail strikes in the 1950s just compounded the problem. Freight traffic was being lost at an alarming rate to road transport and by 1955 BR recorded its first operating loss. More drastic surgery was now required for the ailing railways of Britain.

PREVIOUS SPREAD: Trainspotting was a popular pastime for schoolboys in the 1950s and 1960s. Supported by a plethora of books published by Ian Allan, they spent all their spare time and pocket money standing on draughty station platforms and visiting dark and dirty engine sheds in search of that elusive locomotive number. Here a group of schoolboys watch the comings and goings of steam trains on the East Coast Main Line at Newcastle in August 1950.

FAR LEFT: Before Dr Beeching appeared on the scene, over 3,000 miles of loss-making branch lines were closed by British Railways following nationalization. Here the Branch Line Reinvigoration Society exhorts people on the Isle of Wight to make use of the Newport to Sandown line or it will face closure. This was all to no avail as it was closed on 6 February 1956.

LEFT: British Railways' corporate identity chosen to adorn the sides of locomotives between 1948 and 1956 was the lion rampant, nicknamed the 'Unicycling Lion'.

BELOW: The closures just kept coming – Class 'H' 0-4-4T No. 31518 calls at Brasted on 28 October 1961, the last day of passenger services on the Westerham branch line in Kent.

British Railways standard locomotives

The newly nationalized railways of Britain inherited on 1 January 1948 a massive fleet of steam locomotives, some dating from the period when William Gladstone was still Prime Minister in 1874. Despite the introduction of standardized parts including boilers and driving wheels by the GWR's G.J. Churchward and Charles Collett and their protégé, William Stanier of the LMS, many of Britain's steam locomotives had their origins in a bygone age.

Diesel traction was still in its infancy and it would be several years before suitable designs could be mass-produced in sufficient quantity to replace the tried-and-tested steam locomotive. The progress of steam locomotive development by the 'Big Four' privatized companies in the 1930s had been halted in its tracks by the onset of the Second World War, and by 1948 steam motive power and the entire railway system were in a totally rundown state. Despite this, some proven and ground-breaking designs from the 'Big Four' continued to be built into the British Railways' era – notable among the former were Collett's 'Castle Class 4-6-0s, Stanier's ubiquitous Class '5MT' (Black Five) 4-6-0s and Peppercorn's Class 'A1' 4-6-2s. Of these, Oliver Bulleid's highly innovative 'Merchant Navy' and the lighter 'West Country' and 'Battle of Britain's 4-6-2s, and Hawksworth's 'County' Class 4-6-0s, are noteworthy.

With nationalization looming in 1947, the newly formed Railway Executive appointed Robert Riddles as the first Chief Mechanical & Electrical Engineer of British Railways. Appointed as assistant to the LMS's CME, William Stanier, in 1935, he was closely involved in the design and construction of the 'Coronation' Pacifics, taking part in the record-breaking test run of No. 6220 *Coronation* when the loco recorded a speed of 114 mph south of Crewe. Riddles subsequently accompanied this famous streamlined locomotive on its tour of North America in 1939 and, due to the illness of its regular driver, drove it for much of the time. During the Second World War he was appointed Director of Transportation Equipment for the Ministry of Supply and then Deputy Director General, Royal Engineer Equipment. In this role he was not only involved in the design and

production of Bailey bridges and the D-Day Mulberry Harbours, but also the design and construction of hundreds of the War Department 'Austerity' 2-8-0 and 2-10-0 heavy freight locomotives.

Riddles' new job with BR involved the introduction of a series of new and efficient steam locomotives utilizing interchangeable standardized parts, thus permitting the wholesale withdrawal of outdated steam types and the consequential saving to the British taxpayer. He went on to design twelve varied locomotive classes ranging from the humble Class 2MT 2-6-2Ts right through to the powerful freight '9F' 2-10-0s.

Riddle's standard-class locomotives were by 1951 starting to emerge resplendent from the principal locomotive works around the country – most notable were the 'Britannia' Class 7P6F 4-6-2s, which were soon delivering outstanding performances on the Great Eastern main line from Liverpool Street; the Class 4, Class 5 4-6-0s; and the Class 4 2-6-4Ts. All twelve new classes had been introduced by 1954, culminating in the heavy freight Class '9F' 2-10-0s which mark the epitome of British steam locomotive design. However, one year later the writing was already on the wall for steam traction with the publication of the BR 'Modernisation Plan' – an ill-conceived and ill-timed document, which announced the death sentence for steam and the rapid introduction of, at that time, many untried main-line diesel types.

Steam locomotive construction continued until 18 March 1960 when the final example (and the 999th Standard loco) to be built by BR, Class '9F' 2-10-0 No. 92220 *Evening Star*, emerged from Swindon Works. Just over eight years later BR withdrew from service the last standard-gauge steam locomotives – with a planned 40-year working life many of Riddles' fine locomotives survived less than a decade, with No. 92220's lasting a mere five years.

BELOW: Designed by Robert Riddles, 'Britannia' Class 4-6-2 No. 70000 *Britannia* was the first of 999 standard locomotives built by British Railways between 1951 and 1960. Now preserved, this express passenger locomotive is seen at Neasden shed in January 1951 shortly after appearing new from Crewe Works.

Early diesel, gas turbine and electric locomotives

Even before the Second World War, many American railroads had invested heavily in diesel electric technology as a replacement for steam traction due to problems with coal supplies caused by industrial unrest. Apart from a few faltering attempts in the 1930s, Britain lagged behind until after the Second World War when two of the 'Big Four' railway companies constructed experimental diesel-electric locomotives – the GWR experimented with gas turbine technology but this led nowhere. However, the diesel-electric prototypes paved the way for BR's first generation main-line diesels introduced in the late 1950s.

Gas turbine locomotives

In the 1940s the GWR sought an alternative motive power source in the form of gas turbine technology. The belief was that a gas turbine locomotive could produce power equal to their 'King' Class 4-6-0 whereas two diesel-electric locomotives were needed for the same task. Following the Second World War, the GWR ordered the prototype gas turbine locomotive No. 18000 in 1946. The A1A-A1A loco, built by the Swiss Locomotive Works and Brown, Boveri in Switzerland, was delivered to BR in 1950. Metropolitan-Vickers of Manchester built a slightly more powerful Co-Co loco, No. 18100, delivered one year later. Assigned to main-line services from Paddington, both locos were heavy on fuel and proved unreliable. Withdrawn in 1958, No. 18100 was then used as a test-bed 25kV AC electric loco. In this guise it ran as E1000 (later E2001) until withdrawn from service in 1968. No. 18000 was returned to Europe after its withdrawal in 1960, initially employed as a test bed (minus its gas turbine equipment) by the International Union of Railways and subsequently becoming a static exhibit in Vienna. Later returned to the UK, it is currently on display at Staveley Barrow Hill roundhouse.

Diesel-electric locomotives

The LMS was the first to introduce main-line diesel electric locos after the Second World War. Two Co-Co locos, Nos. 10000 and 10001, designed by H.G. Ivatt, the CME of the LMS, were built at Derby Works, with the first being delivered only two months before Nationalization at the end of 1947. No. 10001 followed later in 1948. Both locos had a power output of 1,600 hp with their English Electric units. Due to their low-power output compared to the 'Coronation' Pacifics, they worked in multiple when entrusted with heavy passenger trains. A connecting corridor was fitted in the front nose of each loco to facilitate this. After use on the Midland and West Coast Main Lines they were transferred to the Southern Region in 1953, where their performance could be compared with the more powerful Bulleid diesel-electric locos Nos. 10201–10203. In 1955 the five locos were permanently transferred to the LMR and allocated to Willesden depot, making regular appearances on the WCML working in multiple

hauling expresses such as 'The Royal Scot'. No. 10000 was withdrawn in 1963 and 10001 was withdrawn in 1966.

Oliver Bulleid designed the Southern Railway's (SR) diesel-electric locomotives but BR actually constructed the slab-fronted mainline diesels. The first two locos (Nos. 10201 and 10202), with English Electric power units and an output of 1,750 hp, emerged in 1950 from Ashford Works and were soon diagrammed to work expresses out of Waterloo. 10000 and 10001 joined them in 1953 for joint performance trials. A third Bulleid-designed loco, No. 10203, with a power output of 2,000 hp, emerged from Brighton works in 1954. Permanent transfer of the five locos to the LMR took place in 1955. Allocated to Willesden, they worked trains on the WCML. All three SR locos were withdrawn in 1963.

Electric locomotives

Before the war the SR's extensive third-rail passenger network radiating south and southwest from London was operated entirely by electric multiple units while freight traffic was entirely steam hauled. In an effort to reduce both maintenance and fuel costs, the SR built two Co-Co electric locomotives at Ashford Works in 1941 and 1945 respectively – the body and bogies for these 1,470-h.p. machines was designed by Oliver Bulleid while a third loco was built by BR in 1948. Numbered 20001–20003 they successfully remained in service until being withdrawn in 1968.

The Second World War had interrupted overhead electrification of the steeply graded LNER route through the Pennines between Manchester and Sheffield (via Woodhead Tunnel). While a prototype electric locomotive, designed by Sir Nigel Gresley, had been built at Doncaster in 1941, electrification of the route was only completed in 1954. Motive power comprised 58 1.5 kV DC Class 'EM1' Bo-Bo mixed-traffic locomotives, all built at Gorton Works between 1950 and 1953 with the exception of the prototype, and seven more powerful Class 'EM2' Co-Co express passenger locomotives, also built at Gorton.

Woodhead route passenger services ceased in 1970, by which time the Class 'EM2' locos had been withdrawn and sold to the Dutch national railway. Freight continued in the hands of the Class 'EM1' until 1981 when the line closed and the locomotives were withdrawn.

LEFT: Designed by George Ivatt for the LMS, 1,600-hp diesel-electric Co-Co No. 10001, seen here at Bletchley on 1 September 1958, was introduced by British Railways in July 1948 and together with sister engine No. 10000 was often employed hauling the 'Royal Scot' between London and Glasgow.

ABOVE LEFT: Experimental gas turbine locomotive No. 18000 is seen here at Bristol Temple Meads station on 31 May 1952 after hauling the 'Merchant Venturer' from Paddington.

ABOVE RIGHT: Designed by Sir Nigel Gresley, 'EM1' Class Bo-Bo electric locomotive No. 26020, seen here on 15 February 1951, was one of fifty-eight built at Doncaster Works for the newly electrified line between Manchester and Sheffield. This locomotive is now preserved at the National Railway Museum in York.

The 1955 'Modernisation Plan'

By the 1950s the nationalized but outdated railways of Britain were struggling to keep their head above the water and by 1955 British Railways had recorded its first operating loss. It was fairly apparent that the whole system required modernizing and overhauling so on 25 January 1955 the British Transport Commission's *Modernisation and Re-equipment of British Railways* was published. Launched at three press conferences chaired by BTC Chairman Sir Brian Robertson, this 35-page 'blueprint' document recommended a £1.24 billion (£25 billion at today's prices) investment over a period of fifteen years to modernize and re-equip British Railways. The document outlined five key areas for modernization:

1. Improvement to track and signalling to allow higher speeds over trunk routes; increased use of colour-light signalling; more power-operated signal boxes; centralised traffic control. £210 million

2. Steam must be replaced as a form of motive power with electric or diesel traction being rapidly introduced; electrification of large mileages of route and the introduction of several thousand electric or diesel locomotives. £345 million

Of the two types of traction, diesel won the day as the cost of implementing widespread electrification was enormous. While diesels were more expensive to build than steam locomotives, their benefits compared to steam were obvious – they were cleaner, their maintenance and running costs were lower and their range was much greater.

3. Replace existing steam-drawn passenger rolling stock, largely by multiple-unit diesel or electric trains; remaining passenger stock must be modernised; modernisation of stations and parcel depots. £285 million

4. Freight services must be drastically remodelled; continuous brakes will be fitted to all freight wagons; faster and smoother operation of freight services; marshalling yards and goods terminals will be resited and modernised, with the number of the former being greatly reduced; larger wagons will be introduced particularly for mineral traffic; loading and unloading appliances will require extensive modernisation. £365 million

5. Improvements at ferry ports, staff welfare, office mechanisms; development and research work associated with the Plan. £35 million

TOTAL £1,240 million
SAY £1,200 million

This was all well intentioned – albeit involving vast amounts of taxpayers' money – but the authors of the 'Modernisation Plan' had failed not only to take into account the almost complete lack of expertise in building diesel locomotives in Britain, but also the selfish and autonomous aspirations of some regions of British Railways – in fact they were still being run more-or-less on the same lines as the old 'Big Four' prior to nationalization.

BELOW: Published on 24 January 1955, the *Modernisation and Re-equipment of British Railways* recommended spending £1.2 billion on Britain's rundown and outdated railways.

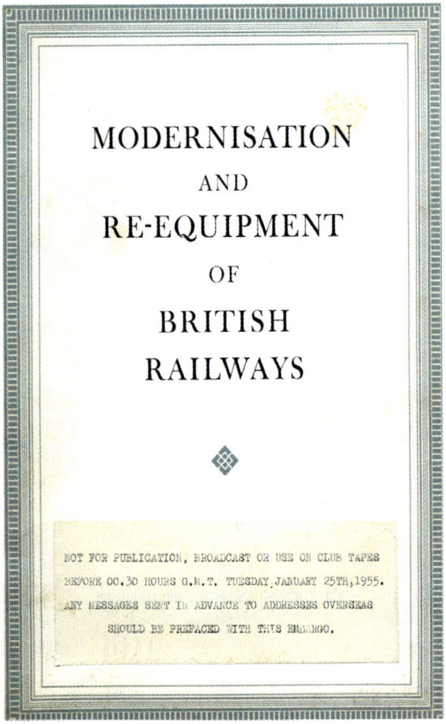

Other fascinating titbits from the 'Modernisation Plan' included:

The use of atomic power in relation to railways seems likely to be indirect, namely through the use of nuclear energy at electric power stations, rather than through the development of atomic-powered locomotives.

It is proposed shortly to bring to an end the building of new steam locomotives, there are present some 19,000 on BR, a substantial proportion being of modern design. The steam locomotive has a useful life in service of some forty years...

...some capital expenditure on steam depots will be inescapable to meet essential requirements.

There will be a substantial expansion in container transport...for the transfer of traffic between road and rail.

...there will be a marked reduction in stopping and branch-line services which are little used by the public and which should be largely handed over to road transport.

So far British Railways have not been given a share of post-war capital investment comparable with that allocated to railways in many other countries.

The 'Modernisation Plan' was backed by the then Conservative government but by 1960 the latter was having second thoughts as BR's losses continued to mount inexorably. By 1963 the partly implemented 'Modernisation Plan' was ditched in favour of Dr Beeching's *Reshaping of British Railways* – by then the political knives were out for Britain's railways and Dr Beeching's 'Axe' was intended to be the *coup de grâce*.

BELOW: Seen here at Liverpool Street station on 18 April 1958, D200 was the first of 200 Type 4 2,000-hp diesel-electric locomotives to be built by English Electric and Robert Stephenson & Hawthorns for British Railways. They had soon replaced steam haulage on the London to Norwich route and remained in service on BR until the early 1980s.

First-generation diesel-electric locomotives

Apart from the five examples built by the LMS and Southern Railway after the war, it took until the publication of the 'Modernisation Plan' in 1955 before British Railways started to invest in diesel-electric locomotives. At that time there were around 19,000 operational steam locomotives on BR's books and the replacement of these by a lesser number of diesel and electric locos was then reckoned on taking around thirty years. The first contracts for these new diesels went to both the private sector – North British Locomotive Company, English Electric and Brush – and to BR's own locomotive workshops but very few of these had had any experience in building such complex and powerful machines.

Despite the lessons to be learnt from steam locomotive design, standardization of diesel locomotives was never part of the master plan and within a few years perfectly good and, in many cases, nearly new steam locomotives with a potential working life of thirty-to-forty years were being sent to the scrap yard to make way for a motley collection of first-generation diesel-hydraulic and diesel-electric locos. By far the biggest culprit was the autonomous Western Region, which decided to opt for lightweight diesel-hydraulic transmission based on a successful German design – the story of this short-lived and unique era is told on the following pages.

Categorized by engine power, five types of mainline diesel locomotives were introduced under the 'Modernisation Plan': Type 1 (800-1,000 hp); Type 2 (1,001-1,499 hp); Type 3 (1,500-1,999 hp; Type 4 (2,000-2,999 hp); Type 5 (over 3,000 hp). Within each of these categories was a mixed bag of designs, some of which were being withdrawn from service before the end of steam in 1968. With hindsight the whole exercise was botched, with enormous amounts of taxpayers' money literally going up in smoke at the breaker's yard.

Opting for heavier electric transmission rather than Swindon's preference for hydraulic transmission, other regions had more success, with the London Midland Region's (LMR) Sulzer-engined Type 4 'Peaks' and the Eastern Region's (ER) English Electric Type 4s and Brush Type 2 being reasonably successful, especially the latter when re-engined, some of which still survive today. Introduced in 1961, the Type 5 'Deltics' were impressive machines but even they had a working life of only twenty years speeding up and down the East Coast Main Line. In fact the most successful design was not a main-line locomotive but the humble diesel shunter (now classified 08), which was introduced in 1953.

BELOW: Introduced in 1961, the 3,300-hp English Electric Type 5 'Deltic' diesel-electric locomotives replaced the streamlined 'A4' 'Pacific' steam locomotives on East Coast Main Line express services. They had all been withdrawn by 1982.

To be fair to Swindon, other regions also had their spectacular failures, notable among them being the Scottish Region's Type 2s (with a life of just eight years), whose chronic unreliability brought about the collapse of their builders, the world-renowned North British Locomotive Company; the same region's Clayton Type 1s (six years); the LMR's Metrovick Type 2s (ten years); and the ER's English Electric Type 2 'Baby Deltics' (eleven years).

Fortunately BR got it right with a few of their first-generation diesels, some of which, such as the English-Electric Type 1s (or Class '20' in current parlance), Type 3s (Class '37') and Brush Type 4s (Class '47's) can still be seen in service today after a working life of fifty years, but by the late 1980s the writing was on the wall for British diesel-electric locomotive manufacturers. Since then Britain has bought its locomotives from the North American companies of General Electric and General Motors and from Vossloh in Spain.

Diesel multiple units

The introduction of the first diesel multiple units by British Railways preceded the 'Modernisation Plan' by several years, with 100 lightweight single-car and two/three/four-car sets being built at Derby Works between 1952 and 1955. With their low operating costs, fast acceleration and cleanliness they were an immediate hit both with BR and with the travelling public. Following this success the 'Modernisation Plan' proposed the building of around 4,500 diesel multiple units and while some of these were built by BR at Swindon and Derby, many were sub-contracted to private sector companies such as Birmingham Railway Carriage & Wagon Co, Gloucester Railway Carriage & Wagon Co, Metropolitan-Cammell, Pressed Steel and Cravens. Eastleigh Works also built a series of diesel-electric multiple units for the Southern Region. Thirty 4-wheel railcars were built by British United Traction, Derby Works, Bristol/Eastern Coach Works, D. Wickham & Co., Park Royal and AC Cars, while five were supplied by Waggon und Maschinenbau of West Germany. The unique 2-car battery electric unit, nicknamed 'The Sputnik' and used on the Dee Valley line west of Aberdeen, was jointly built by BR at Derby and Cowlairs Works.

The publication of the 'Beeching Report' in 1963 and the subsequent closure of around 4,000 miles of railway put an end to further first-generation diesel multiple unit construction.

Birmingham Railway Carriage & Wagon Co. and Derby Lightweight diesel multiple units recreating an early 1960s scene at Cheddleton on the Churnet Valley Railway in 2008.

Diesel-hydraulic locomotives

Before the 'Modernisation Plan' was published in 1955 there had been very little research and development carried out by British Railways with regard to diesel traction. The only exceptions were the two LMS-designed diesel-electric locomotives Nos. 10000 and 10001 and the three Southern Railway-designed Nos. 10201–10203. While the 'Plan' forecast the end of steam it was envisaged that this would happen gradually over a period of thirty years, as many of BR's steam locomotives were either new or would continue to be built for a few more years. In reality, within a short space of time the British Transport Commission had hurriedly ordered nearly 3,000 main-line diesel locomotives of various non-standard types even before prototype testing had been properly evaluated. Most of these locomotives were diesel-electrics, as were the LMS and SR prototypes, where a diesel engine drives an electric generator that in turn provides power to drive electric motors connected to the wheels. However, the semi-autonomous Western Region at Swindon thought differently.

The Western Region (WR) opted for diesel-hydraulic transmission, in which the power from a diesel engine is transmitted to the wheels via hydraulic transmission through a torque converter. Their decision was based on the success of German lightweight V200 diesel-hydraulic locomotives that had been in service on West German railways since 1953 – the WR placed orders in 1956 to build a British version using German engines and transmission made under licence.

The first of the WR diesel-hydraulics to enter service were five Type 4 A1A-A1A 'Warships', numbered D600–D604, built by the North British Locomotive Company (NBL) in Glasgow. They were delivered between early 1958 and early 1959 and were employed on main-line expresses between Paddington and Plymouth but their poor reliability, high maintenance costs and non-standard design led to their mass withdrawal in late 1967.

Two batches of Type 4 B-B 'Warships' followed – D800–D832 and D866–D870 were built at Swindon Works with deliveries commencing in 1958, and D833–D865 were built by NBL and delivered from 1960 onwards. Like their predecessors both batches also suffered from high running costs and poor reliability despite replacing tried and tested steam haulage on main-line routes from Paddington to the West Country and Bristol. Withdrawals took place between 1968 and 1972.

The third type of diesel-hydraulic were the Type 2 B-B 'Baby Warships', numbered D6300–D6357, which were also built by NBL and entered service on secondary and branch-line duties between 1959 and 1962. Initially they were plagued with engine and transmission problems but these were rectified by the manufacturers and they

then gave a reasonably reliable service. Withdrawals commenced in 1967, by which time NBL had gone bankrupt, and the remaining fleet was kept going by using cannibalized spares from withdrawn locos. By early 1972 the class had become extinct.

The fourth type was certainly much more successful than previous WR diesel-hydraulics. Introduced in 1961 the Type 3 B-B 'Hymeks', numbered D7000–D7100, were built by Beyer-Peacock in Manchester. After initial problems the stylish locomotives proved to be very reliable and could be seen across the WR network on a wide range of duties. Withdrawals took place between 1971 and 1975.

The final batch of main-line diesel-hydraulics were the equally stylish and powerful Type 4 C-C 'Westerns', numbered D1000–D1073, which were introduced between late 1961 and 1964. They were built in two batches, D1000–D1036 at Swindon Works and the remainder at Crewe Works, and were soon in charge of the WR's expresses between Paddington and the West Country, South Wales and Birmingham. Fitted with twin Maybach engines they suffered from transmission and train heating problems, and were withdrawn between 1973 and 1977, their place being taken by the highly successful new Inter-City 125s, which remain in service to this day.

The final class of diesel-hydraulics were the Type 1 0-6-0 shunting and short-trip locomotives, numbered D9500–D9555, which were built at Swindon Works between 1964 and 1965. Of all the hydraulic designs the 'Teddy Bears', as they were nicknamed, had the shortest working life on the WR, with withdrawals taking place between 1968 and the end of 1970. Some escaped the cutters' torch and found new working lives on industrial railways.

And so ended the Western Region's brief, colourful and expensive experiment with diesel-hydraulics. Their lack of compatibility with diesel-electric traction in the other regions coupled with generally poor reliability and high maintenance costs led to their rapid fall from grace. Autonomy at Swindon had finally ended.

LEFT: North British 'Warship' Type 4 diesel-hydraulic D857 *Undaunted* heads the 13:25 Newquay to Paddington express through Dawlish, circa 1962. When withdrawn on 3 October 1971 it was the last NBL 'Warship' in service and was simply switched off when still in full working order.

ABOVE: On 20 July 1973, Beyer Peacock 'Hymek' Class '35' No. (D)7001 heads a train of empty bitumen tanks away from Cranmore station sidings, now the home of the East Somerset Railway. This locomotive was withdrawn from service during March 1974.

The trainspotting craze

During the 1950s and 1960s it was a common sight to see a gaggle of young lads standing at the end of platforms on busy stations around the country – notebooks, pencil and possibly a camera in their hands, they were often dressed in duffle coats and usually carried a duffle bag containing sandwiches and drinks. They were trainspotters. Passengers on trains also looked on in alarm as this strange breed rushed excitedly up and down the corridor coaches, craning their heads out of windows to catch a glimpse of locomotives in engine sheds and sidings.

Writing down the numbers of steam, and later diesel, locomotives was not a new phenomenon in the post-war years – the 'Big Four' had previously published books containing lists of their own locomotives *'specially arranged for engine spotters'*. However, the hobby really took off when a young Southern Railway clerk by the name of Ian Allan published the *ABC of Southern Locomotives* in 1942 – price one shilling, the book was a big hit and subsequently led to a whole series of regional ABC Guides, Locomotive Shed Directories, Loco Shed books and Locolog books along with the monthly railway magazine *Trains Illustrated*. By the 1950s trainspotting had become an all-absorbing hobby for thousands of boys around Britain who spent all their spare time (and cash) chasing after rare and elusive locomotive numbers.

The most treasured possession of a trainspotter was Ian Allan's *Combined Volume of all British Railways' Locomotives* which, in the Summer of 1961, cost 10*s*. 6*d*. (new money 52½p), a vast amount to shell out for an impoverished schoolboy! Once a locomotive number had been spotted it was neatly underlined in the book but very few ever got to spot every engine as this would entail journeys of hundreds of miles to some far away engine shed – there were 19,000 steam locomotives on BR in 1955. Railway societies sprouted up all over the country to ease this burden, arranging motor coach trips on Sundays when most locomotives were tucked up in their sheds. The author well remembers one such trip organized by the Warwickshire Railway Society on Sunday 7 April 1963 when a coachload of sleepy trainspotters travelled down from the Midlands to South Wales – in just this one day we visited no less than twenty-one steam engine sheds and spotted 619 locomotives! In trainspotting terms this was known as a 'shed bash' and spotting a new number was a 'cop'. On organized visits like these shed permits were essential but many trainspotters resorted to illegal means by sneaking round engine sheds out of sight of the foreman – in trainspotting terms this was called 'bunking'. Health & Safety did not seem too important then. Railway locomotive works such as Swindon, Crewe, Derby and

Eastleigh also had rich pickings for trainspotters but guided tours of these were normally confined to Sundays when the workforce was absent.

The hobby wasn't just confined to collecting numbers either. Various organizations such as the Branch Line Society, the Stephenson Locomotive Society and the Locomotive Club of Great Britain arranged special trips along freight-only lines or railways that were about to close – society members were often conveyed along these weed-infested tracks in antiquated coaches hauled by veteran steam locomotives.

Although steam haulage was doomed by the early 1960s the trainspotting craze continued unabated with new diesel locomotives fresh out of works adding colour to the already varied scene. Even after steam had disappeared from BR in 1968 the trainspotting craze continued but in more recent years the gradual replacement of locomotive-hauled trains by bland multiple units, health and safety regulations and security concerns by the railway authorities have all conspired to bring about its virtual downfall. There are still some hardened enthusiasts out there with notebooks in their hands but the trainspotting phenomenon of fifty years ago is now just a fading memory.

LEFT: A school party visits Basingstoke depot on Friday 2 June 1967 during the last week of steam on the Southern Region. Rebuilt West Country 4-6-2 No. 34004 *Yeovil* is lined up alongside future traction, an electro-diesel (later class 73), and Brush Type '4' (later class 47) D1848.

BELOW: A page from one of the author's many trainspotting notebooks, detailing locomotives that he spotted at Polmadie shed in Glasgow on 29 March 1964. Judging from the number of ticks (cops) it was a good day for Julian!

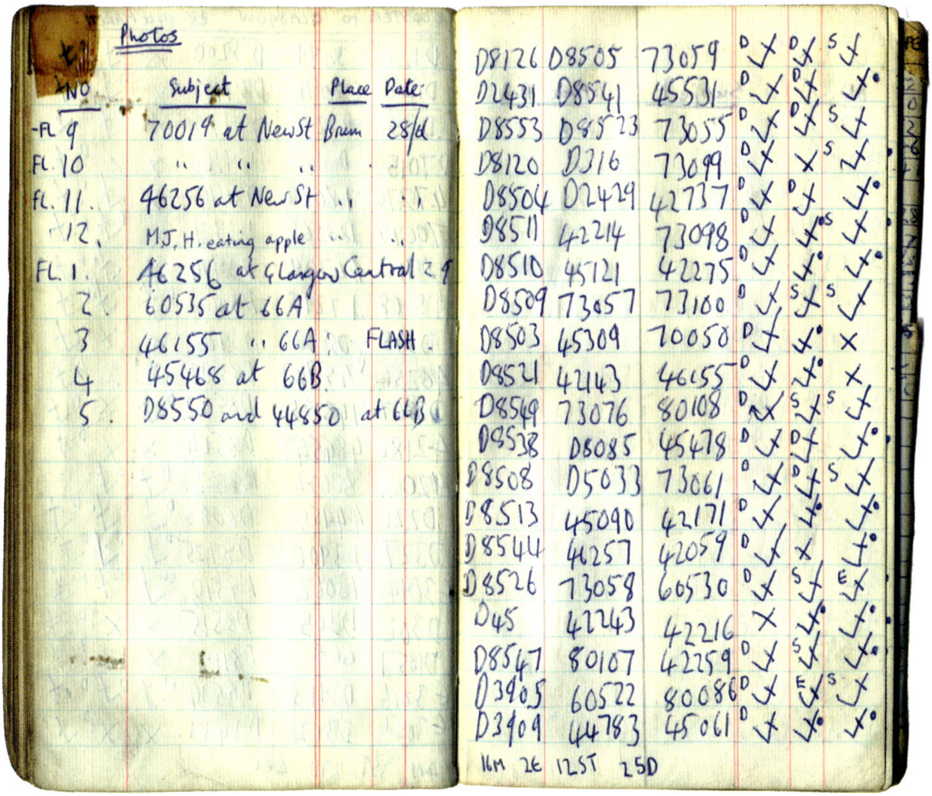

Post-war electrification

As we have already read in the previous chapter, the only major electrification scheme on Britain's railways before the Second World War was the expansion of the third-rail network by the Southern Railway from London to the south coast. Although Brighton and Portsmouth had already been reached it took until after the war for further electrification to be resumed. The Kent Coast routes were electrified by 1961 and at the same time the voltage across the whole of the BR (Southern Region) network was increased from 600V DC to 750V DC. The South Western Main Line from Waterloo to Southampton and Bournemouth was electrified in 1967 and extended to Weymouth in 1988. Other routes electrified during the 1980s were the Hastings Line, East Grinstead branch and Eastleigh to Fareham.

In the north of England the electrification of the LNER's 41½-mile Woodhead Line between Manchester and Sheffield was put on hold during the war, finally being completed in 1955. Using overhead wires for power collection, the line was electrified at 1.5kV DC which within a few years made it obsolete as further electrification on Britain's main lines was then standardized at 25kV AC.

Although the 'Modernisation Plan' of 1955 promised the completion of major electrification schemes on Britain's railways, in reality progress was extremely slow – electrification was an expensive business and, in the straitened times in which Britain found itself, Government funding was not forthcoming. Instead an interregnum period of dieselization preceded many of these schemes. The first major main-line electrification took place on

the West Coast Main Line between London Euston and Glasgow Central, consequently disrupting Britain's busiest rail route over a period of sixteen years before its completion in 1974.

Electrification of suburban lines around Glasgow started in 1960 with the introduction of the new 'Blue Trains' on the North Clyde Line and since then has been extended on many other routes, but fifty-five years later it is still incomplete. Suburban lines out of London's Liverpool Street and Fenchurch Street stations were also electrified in the 1960s but it took many years before the Great Eastern Main Line was complete – Colchester was reached in 1961 but it took until 1987 before the entire route to Norwich was 'switched on'. The line from Liverpool Street to Cambridge and King's Lynn was electrified in stages starting in the 1960s but only completed in 1992.

Once the stamping ground of Gresley's 'A4' Pacifics and their successors, the powerful 'Deltic' diesels, the East Coast Main Line was electrified in two distinct stages – the southern end from King's Cross to Royston was completed in 1978 but it took until 1991 for the entire route to Leeds and Edinburgh to be complete.

Elsewhere, two commuter lines operating out of Birmingham New Street were electrified in 1966 and 1993 respectively, while local lines around Leeds were 'switched on' in the 1990s. Back in London local services from St Pancras to Bedford were electrified in 1983 and the Heathrow Express from Paddington in 1994.

Sadly, despite the obvious long-term benefits, electrification of Britain's railways has been carried out in a piecemeal fashion since the 1950s and in 2007 the then Labour government only made matters worse by ruling out any large-scale schemes for the next five years – in the same year the electrified High Speed 1 link between the Channel Tunnel and St Pancras became the first main line to open in Britain for over 100 years.

The coalition government announced in 2012 a major electrification programme. Routes earmarked were the Northern Hub in North West England and Yorkshire, the Midland Main Line from Bedford to Leicester, Derby and Sheffield and the Great Western Main Line from London Paddington to Bristol, Cardiff and Swansea along with the South Wales Valleys' routes. However, soon after taking office following the May 2015 general election, the new Conservative government announced the postponement of all of these plans apart from the Great Western Main Line out of Paddington on which work had already started. The latter is due for partial completion in 2017 but there already delays – in fact the only electrification scheme on schedule is Crossrail across London which is due to be completed in 2018. Work has just started on electrifying the Edinburgh to Glasgow route with completion due for 2019.

Britain's railways are still suffering from a lack of investment and many other important routes will be waiting for many years before they are considered for the electrification treatment. Notable among these are the North Wales Coast route from Crewe to Holyhead, the Midland main line between Bristol and the Midlands and routes to Devon and Cornwall.

FAR LEFT: Poster produced in 1961 for the London Midland Region's electrification scheme when electric trains were introduced to the Liverpool to Crewe section of the West Coast Main Line.

LEFT: From an original painting by Terence Cuneo, this 1960s poster was produced for the Scottish Region to advertise the new Glasgow electric 'Blue Trains' that had been introduced on the North Clyde line to Helensburgh.

Railway corporate design and publicity

The post-war years brought nationalization of Britain's railways and a new period of corporate identity. Between 1948 and 1956 the lion rampant (nicknamed the 'Unicycling Lion') was chosen to adorn the sides of locomotives while a hot-dog sausage-shaped totem was selected to appear on BR publicity material and on station nameboards, colour coded for each of the six regions. With some variations – the 'ferret and dartboard' crest replaced the lion rampant in 1956 – these remained as the BR corporate identity until the introduction of the famous double-arrow logo in 1965.

By the early 1950s the newly formed British Railways' publicity department had got back into full swing after the post-war austerity period and were producing posters, holiday guides, brochures, leaflets and regional timetables, all carrying the corporate identity logo. Numerous named trains, some reintroduced after the war but many brand new, received the full publicity treatment with personalized locomotive headboards and matching leaflets and restaurant car menus – the age of romantic rail travel appeared, at least on the surface, to have returned.

LEFT: From an original painting by Frank Wootton, this poster features a BR Standard Class 5MT 4-6-0 'somewhere in Wales' and was produced by the Western Region in 1960 to promote rail travel to the principality.

BELOW: From an original painting by Reginald Mayes, this 1950s poster was produced by the Eastern Region to promote its 'The Queen of Scots' Pullman service between London and Glasgow.

Terence Cuneo

Born in London in 1907, Terence Tenison Cuneo was without doubt the finest poster artist of the British Railways era. After studying at the Slade School of Art he started his career as a magazine and book illustrator before being appointed as an official war artist during the Second World War, painting scenes of aircraft factories and wartime activities. He continued his commercial work after the war, specializing in painting railway scenes before being appointed official artist of Queen Elizabeth II's Coronation in 1953. Cuneo then went on to be commissioned by British Railways to produce a large number of oil paintings of the railways, their locomotives and infrastructure. Often putting his own personal safety at risk when sketching on location (for example perching high on the girders of the Forth railway bridge), Cuneo's masterpieces always included his personal trademark – a small playful mouse hidden away in a corner of each painting. His output continued into the diesel and electric era, and today his posters are prized and valuable possessions for collectors of railway ephemera. Cuneo died in 1996 and is immortalized, complete with paintbrush, palette and trademark tiny mouse, by Philip Jackson's larger-than-life bronze statue at Waterloo station in London.

BELOW: Terence Cuneo often put his life in danger when sketching his railway subjects. In order to produce this painting for a BR poster he tied himself to one of the high girders on the Forth Bridge, having to withstand biting 50 mph winds suspended above the track.

Famous named trains

As we have read previously, the 'Big Four' railway companies all operated named trains in the 1920s and 1930s but the outbreak of war in 1939 brought an immediate cessation of most of these services. The austerity of the immediate post-war years was followed in 1948 by the nationalization of Britain's railways and it was not long before the newly formed British Railways not only started to reintroduce the pre-war named trains but also a large number of new ones. Several of BR's regions stamped their own style on accompanying publicity material such as fold-out route guides, luggage labels and restaurant car menus. Locomotive headboards added individuality to the trains and the Western Region's ornate examples, along with brown and cream coaches, certainly made them stand out from the crowd.

Naming trains was a short-lived affair for British Railways and by the 1960s, with steam on the way out, these trains had somewhat lost their glamour – the new diesel locomotives at their head soon lost their nameboards and by the end of that decade most had ceased to exist.

The decade of the 1950s was certainly a new, but brief, golden age for these named trains with the following new services added to the reintroduced pre-war services:

Western Region (WR)

'The Inter-City' (1950–1965)
London Paddington to Birmingham Snow Hill/ Wolverhampton Low Level

'The Red Dragon' (1950–1965)
London Paddington to Carmarthen

'The Merchant Venturer' (1951–1965)
London Paddington to Weston-super-Mare

'The Cornishman' (1952–1962)
Wolverhampton Low Level to Kingswear/Penzance

'Pembroke Coast Express' (1953–1963)
London Paddington to Pembroke Dock

'The Royal Duchy' (1955–1965)
London Paddington to Kingswar/Penzance

'The South Wales Pullman' (1955–1961)
London Paddington to Swansea

'Capitals United Express' (1956–1965)
London Paddington to Cardiff
'Cathedrals Express' (1957–1965)
London Paddington to Worcester and Hereford
'The Mayflower' (1957–1965)
London Paddington to Kingswear/Plymouth

London Midland Region (LMR)
'The Midlander' (1950–1963)
London Euston to Wolverhampton High Level
'The Red Rose' (1951–1966)
London Euston to Liverpool Lime Street
'The Emerald Isle Express' (1954–1975)
London Euston to Holyhead
'The Shamrock' (1954–1966)
London Euston to Liverpool Lime Street
'The Robin Hood' (1959–1962)
London St Pancras to Nottingham

Combined LMR and Scottish Region (ScR)
'The Northern Irishman' (1952–1966)
London Euston to Stranraer Harbour

'The Caledonian' (1957–1964)
London Euston to Glasgow Central

Southern Region (SR)
'The Devon Belle' (1947–1954)
London Waterloo to Ilfracombe/Plymouth (until 1950)
'The Kentish Belle' (1951–1958)
London Victoria to Ramsgate/Canterbury
'The Royal Wessex' (1951–1967)
London Waterloo to Bournemouth/Swanage/Weymouth
'The Man of Kent' (1953–1961)
London Charing Cross to Folkestone

Eastern Region (ER)
'The Day Continental' (1947–1987)
London Liverpool Street to Harwich Parkeston Quay

BELOW: Introduced in 1957 between Hereford, Worcester and London, the 'Cathedrals Express' became the last steam-hauled named train to operate out of Paddington station. It is seen here passing Reading behind 'Castle' Class 4-6-0 No. 7027 *Thornbury Castle* in the late 1950s.

'The Master Cutler' (1947–1968)
London Marylebone/King's Cross to Sheffield
'The Norfolkman' (1948–1962)
London Liverpool Street to Norwich/Cromer
'The South Yorkshireman' (1948–1960)
London Marylebone to Bradford Exchange
'The Fenman' (1949–1968)
London Liverpool Street to Bury St Edmunds (until 1952)/ Wisbech (1953–1960)/Hunstanton (until 1960)/ King's Lynn
'Thc Broadsman' (1950–1962)
London Liverpool Street to Norwich, Cromer and Sheringham
'The Easterling' (1950–1958)
London Liverpool Street to Lowestoft/Yarmouth
'Sheffield Pullman' (1958–1968)
London King's Cross to Sheffield

Combined North Eastern Region (NER) and ER
'Tees-Tyne Pullman' (1948–1976)
London King's Cross to Newcastle
'The Northumbrian' (1949–1964)
London King's Cross to Newcastle-upon-Tyne
'The West Riding' (1949–1967)
London King's Cross to Leeds/Bradford
'The White Rose' (1949–1967)
London King's Cross to Leeds/Bradford

Combined NER, Scottish Region (ScR)
'The North Briton' (1949–1968)
Leeds to Edinburgh Waverley/Glasgow Queen Street

Combined ER, NER, ScR
'The Capitals Limited' (1949–1953)
London King's Cross to Edinburgh Waverley
'The Heart of Midlothian' (1951–1968)
London King's Cross to Edinburgh Waverley
'The Elizabethan' (1953–1962)
London King's Cross to Edinburgh Waverley
'The Fair Maid' (1957–1958)
London King's Cross to Perth
'The Talisman' (1956–1968)
London King's Cross to Edinburgh Waverley

BELOW: Introduced in 1953, 'The Elizabethan' was the crack post-war express between London King's Cross and Edinburgh. Running only in the summer months, it was then the longest scheduled non-stop service in the world. It is seen here in on 8 September 1961 waiting to depart from Edinburgh Waverley behind Class 'A4' 4-6-2 No. 60009 *Union Of South Africa.*

'Blue Pullmans'

Three years after the acquisition of the British Pullman Car Company by the British Transport Commission in 1954, British Railways announced that new luxury high-speed diesel multiple units would be introduced on services between London Paddington, Bristol and South Wales, and between St Pancras and Manchester Central. Built by Metropolitan-Cammell of Birmingham the five new Pullman sets were finished in a striking blue and white and were fitted with air-conditioning and double glazing, with the streamlined power cars at each end fitted with 1,000-hp diesel-electric power units. Providing an at-table service of meals and drinks the on-train staff were smartly dressed in matching blue uniforms. Two 6-car sets were ordered for the London Midland Region and three 8-car sets for the Western Region (WR). Designed primarily for businessmen the 'Blue Pullmans' first entered service on the Midland mainline in July 1960 and prove to be very popular. This service was withdrawn in 1966 when the West Coast Main Line electrification between Euston and Manchester Piccadilly was completed and the two sets transferred to the WR.

The WR 'Blue Pullmans' began service between Paddington and Bristol in September 1960, and an extra service to Swansea was added the following summer. With the additional two sets transferred from the LMR in 1966, an additional service to Bristol and a new one to Oxford was added. The introduction of Inter-City 125s led to their withdrawal in 1973.

BELOW: First introduced on 4 July 1960 between London (St Pancras) and Manchester (Central), the two 6-car 'Midland Pullman' diesel-electric multiple units had a top speed of 90 mph. The train was withdrawn in 1966 on completion of the West Coast Main Line electrification.

Engine sheds

From the very early years of railways the engine shed, or motive power depot, played a vital part in the day-to-day running of goods and passenger trains. Here locomotives were serviced, watered and replenished with coal before leaving for their day's duties. At major engine sheds repair work was also carried out thus saving time that would normally be spent at the railway's main locomotive works. Over a period of time many of these engine sheds had grown into enormous facilities, with some providing home for several hundred locomotives – by the early 1950s Stratford shed in East London had an allocation of nearly 400 and Doncaster boasted nearly 200. In the 1950s and 1960s these palaces of steam had become places of pilgrimage for thousands of young trainspotters, many of whom illegally skulked around the sheds and yards recording the number of each locomotive. They were pretty filthy and dangerous places but to these enthusiasts they were pure heaven!

The architectural style of engine sheds ranged from the simple one-road shed with a pitched roof to multiple-road sheds with north-light pattern or louvre roofs. Based on North American practice, roundhouse sheds were particularly favoured by the North Eastern and Midland railways in the nineteenth century, where locomotives were parked on sidings that radiated out from a central turntable – the only survivors of this type today are the London & Birmingham Railway's roundhouse at Chalk Farm in North London and the Midland Railway's shed at Barrow Hill in Derbyshire. A few new flat-roof concrete engine sheds were built by British Railways in the 1950s but these were deigned to accommodate both steam and diesel locomotives. The larger engine sheds were also fitted with locomotive turntables and enormous concrete coaling towers – the only survivors can be seen at Carnforth shed, now headquarters of the West Coast Railway Company and at Immingham.

Locomotives were allocated to specific engine sheds where they always returned after their turns of duty, sometime spending much of their long working lives based in one place and only being absent when a major overhaul was needed. At the top link sheds such as King's Cross or Old Oak Common, where locomotives were

prepared before hauling a crack express, great pride was taken by staff in cleaning these steam monsters until their paintwork gleamed.

Engine sheds were each given a code, for instance Willesden shed in North London was 1A and locomotives allocated to that shed carried an oval cast-iron plate with the shed code fixed to the bottom of the smokebox door. Smaller engine sheds, or sub-sheds, were not given separate codes as the locomotives that used them were included in the allocation of the mother shed – Anstruther sub-shed in Fife came under the control of nearby 62A Thornton shed.

During the early years of BR modernization, new diesel locomotives were forced to rub shoulders with their older and dirtier steam shed mates but this was not an ideal situation and by the early 1960s some former steam sheds, such as 64B Haymarket, 82A Bristol Bath Road and 88A Cardiff Canton, were being rebuilt as modern diesel maintenance depots.

Despite the influx of diesels there were still thousands of steam locomotives at work on British Railways in the early 1960s. To put this into context, by the summer of 1961 there were just a handful of dedicated diesel depots around the country but nearly 300 of the older steam-age sheds in use and this figure does not include the 147 sub-sheds. However, by 1968 there were just a handful left, all in northwest England, such as Rose Grove, Lostock Hall and Carnforth, and even these closed at the end of steam haulage at midnight on 11 August of that year.

Today's modern diesel-electric and electric locomotives are serviced and maintained at a handful of traction maintenance depots such as Toton in Nottinghamshire and Carlisle Kingmoor in Cumbria (both for diesels) and Crewe (for electrics), or when not in service are stored at stabling points.

ABOVE: The silent interior of the former GWR engine shed at Aberdare (88J) on 15 September 1963. Ranged around the turntable are from left to right 2-8-0 No. 3866, 0-6-0PTs No. 9671 and No. 8723, 0-6-2T No. 5647, 0-6-0PTs No. 3603 and unidentified and 2-8-0T No. 5237. The occupants seen here were soon to be destined for the scrapyard and the shed closed in 1965.

Travelling Post Offices

The Liverpool & Manchester Railway was the first to carry mail by train just two months after it had opened in September 1830. With the opening of the Grand Junction Railway (GJR) to Birmingham in 1837 and the London & Birmingham Railway in 1838 it was possible to send letters from Liverpool or Manchester to London in less than a day, a vast improvement on the previous mail coach schedules. The GJR also introduced the first mail sorting carriage in 1838 by converting a railway horsebox, which allowed the contents of mail bags to be sorted en route. Mailbags on the train could also be exchanged with incoming mail bags via a lineside apparatus while the train was travelling at speed – this ingenious device which could be used while the train was travelling at 60 mph was invented by John Ramsay, a Post Office official. The Railways (Conveyance of Mails) Act of 1838 also required railway companies to carry mail either on scheduled trains or special ones at times directed by the Postmaster-General.

Mail coaches were normally attached to the fastest trains over each route but as the postal business grew it became necessary to introduce dedicated mail trains, the first of which was operated by the Great Western Railway between London and Bristol in 1855. These trains became known as Travelling Post Offices, where mail was sorted on the moving train and could be exchanged either via the lineside apparatus without stopping or at junction stations between other mail trains – a post box was also attached to the side of the sorting carriage so that letters could be posted at stations.

Probably the most important Travelling Post Office (TPO) in Britain was the 'Special Mail', which was introduced in 1885 between London Euston and Aberdeen. Running via the West Coast Main Line the train was operated jointly by London & North Western Railway and the Caledonian Railway, and included three sorting carriages and three mail tender vans plus limited passenger accommodation. This train was the forerunner of the famous 'West Coast Postal', which by the 1930s had grown to six sorting coaches fitted with exchange apparatus and seven parcels vans. Running in both directions the two trains were officially known as the 'North Western TPO Night Down' and the 'North Western TPO Night Up'. On board each train were thirty-five mail sorters and two porters who worked hard through the night sorting mail that had been collected and depositing it via exchange apparatus en route. The 'Postal' also stopped at Rugby, Tamworth, Crewe, Preston, Carlisle, Carstairs, Stirling and Perth, and carried through mail coaches for Glasgow and Edinburgh. The train was

immortalized in the famous documentary film 'Night Mail', which was made by the GPO Film Unit and featured a poem by W.H. Auden and music by Benjamin Britten – the locomotive was LMS 4-6-0 No. 6115 *Scots Guardsman.* Back in the 1930s 80 per cent of all mail in Britain was transported on trains – this amounted to 4.8 billion letters and 140 million parcels each year.

The post-war years saw the number of TPOs travelling around Britain at night drop considerably and the trackside exchange apparatus was last used in 1971 – by 1988 there were thirty-five TPOs left and by 2003 this had dropped to just eighteen. Despite much mail traffic being lost to road and air transport mail trains continued to operate into the twenty-first century. Following privatization of the railways in 1994 the TPOs were operated firstly by Rail Express Systems and then by English Welsh & Scottish Railway. Sixteen sets of new Class '325' 4-car dual-voltage parcel trains were also introduced in 1996 but in 2003 the Royal Mail announced that it was no longer going to use rail to transport mail. The last TPO ran on 9 January 2004 but parcels traffic is still conveyed in the dedicated Class 325 train sets between the three main distribution depots at Willesden in North London, Warrington in Lancashire and Shieldmuir, south of Glasgow.

LEFT: Rail Express Systems diesel-electric locomotive No. 47768 *Resonant* at the head of the 22:25 St.Pancras to Newcastle Travelling Post Office (TPO) shortly before termination of this TPO service on 26 May 1995.

BELOW LEFT: Mail being sorted in the North Eastern Travelling Post Office sorting van in 1987. The last TPO ran on 9 January 2004.

BELOW: The famous 'West Coast Postal' picking up mail from a lineside apparatus while at speed near Harrow on 21 June 1957. Trackside exchange apparatus like this was last used on Britain's railways in 1971.

Rural and seaside branch lines

The opening of railways in rural areas of Britain in the nineteenth century had brought enormous benefits to the scattered communities they served. Farmers were able to send their fresh produce and livestock to distant markets for the first time and country stations provided a previously unheard-of link to the outside world. By the early twentieth century whole trainloads of seasonal produce such as broccoli and potatoes were making their way to Covent Garden Market in London. Milk was also an important source of traffic with milk churns being collected at wayside stations for delivery to creameries, and by the 1930s dedicated milk tanker trains were travelling overnight from the West Country to bottling plants in London.

Along Britain's coastline small harbour towns also benefited greatly from the coming of the railways, with fishermen being able to transport their day's catch overnight in refrigerated vans to the fish markets of the big cities – entire refrigerated trainloads of fish were once despatched from the fishing ports of Fraserburgh and Peterhead in Scotland and Grimsby in Lincolnshire for auction next day at Billingsgate Market in London. Fresh produce from every corner of Britain found its way on to dining tables all thanks to the railways.

The coming of the railways to Britain's coastal towns and villages also led to a boom in home-grown tourism that the railway companys' publicity departments were eager to promote, transforming previously sleepy harbour villages into thriving holiday resorts. The Great Western Railway excelled at this, promoting their route to the West Country as 'The Holiday Line' and naming their premier express from London Paddington to Cornwall the 'Cornish Riviera Limited'. Introduced in 1904 this famous train eventually consisted of eight portions with through coaches for Penzance, St Ives, Falmouth, Newquay and Kingsbridge plus coaches for Ilfracombe and Minehead slipped at Taunton. During Summer Saturdays in the 1930s such was the demand that extra relief trains were laid on.

Railways reached North Devon and North Cornwall in the later nineteenth century with the branch to Padstow, the furthermost destination on the London & South Western Railway's (LSWR) so-called 'Withered Arm', finally being opened in 1899. The LSWR's successor, the Southern Railway, made the most of its West Country destinations by introducing the 'Atlantic Coast Express' in 1927. Apart from the war years the train continued to run until 1964 and during its peak in the 1950s was run on Summer Saturdays as five separate trains from London Waterloo with through coaches to Ilfracombe, Torrington, Padstow, Bude, Seaton, Lyme Regis and Plymouth.

In South Wales the resorts of Porthcawl and Barry Island, both served by short branch lines, were a Mecca for thousands of coal miners and their families, who were transported down from the Valleys in excursion trains on Bank Holidays. Porthcawl became so popular that the GWR opened an enlarged station with extra platforms and carriage sidings in 1916 and by the late 1930s around

BELOW: From an original painting by J. Greenup, this 1930s poster was produced for the LNER to promote train services to Saltburn and Marske-by-the-Sea. Fortunately Saltburn, with its many attractions, is still rail-served today.

70,000 day trippers were arriving by train each week during the summer months.

Sadly by the time of the 1963 'Beeching Report' British holidaymakers were changing their habits. The train ride to the seaside for the annual holiday was becoming a thing of the past and with more expendable income the British were becoming a nation of car owners, quite happy to sit for hours nose-to-tail on the A303, A30 and A38 routes down to the West Country. Cheap holidays in Spain were also taking off thanks to the likes of Freddie Laker and with Dr Beeching's 'Axe' waiting to strike the future looked grim for many rural and seaside branch lines.

While many rural and seaside branch lines did close in the 1960s and 1970s there were some fortunate survivors which are still open for business today. Primary among these are the former GWR branch lines in Cornwall serving St Ives, Falmouth, Newquay and Looe, all of which still offer a stress-free and delightful journey to these popular seaside resorts. North Cornwall and North Devon were not so lucky as the lines to Padstow, Bude and Ilfracombe have long since closed – the only railhead of Barnstaple is now at the end of a long single line from Exeter. In East Devon the branch line to Exmouth escaped closure but those to Budleigh Salterton, Sidmouth, Seaton and Lyme Regis all got the chop.

Elsewhere around Britain's long coastline many other seaside branch lines were also closed in the post-Beeching era – Norfolk was particularly hard hit with lines to Mundesley-on-Sea, Wells-next-the-Sea and Hunstanton all closing. Fortunately a few still cling to life in the twenty-first century and are well worth visiting today. These include Minehead in Somerset (West Somerset Railway), Lymington in Hampshire, Cromer and Sheringham in Norfolk, Skegness and Cleethorpes in Lincolnshire, Whitby and Saltburn-on-Sea in Yorkshire and North Berwick, Largs and Wemyss Bay in Scotland.

Camping coaches

First introduced by the LNER in 1933, self-catering railway camping coaches became popular with British holidaymakers and hikers in the years immediately preceding and following the Second World War. The accommodation was provided in older carriages, suitably converted to provide sleeping, cooking and living space, usually parked in a siding adjacent to a rural or seaside station. Over the next two years the rest of the 'Big Four' railway companies followed suit and by the outbreak of war there were around 100 of these coaches dotted around the network. Revived after the war, camping coaches became popular again with large numbers parked at beauty spots in Scotland and the West Country. As holiday habits changed decline set in and the last camping coaches were withdrawn in 1971. There has been a revival in recent years with fifteen now hired out by railway heritage companies and private owners.

RIGHT: A happy family relax in their Pullman camping coach at Betws-y-Coed on the Conwy Valley line in North Wales on 28 July 1960.

Railway accidents

The first acknowledged railway accident to a passenger occurred on 15 September 1830 when Robert Stephenson's *Rocket* ran over politician William Huskisson at Parkside station during the opening ceremony of the Liverpool & Manchester Railway – he later died of his injuries in hospital. Early railway safety was either non-existent or at best extremely basic, with collisions, runaway trains, excessive speed, derailments, mechanical failure and human error all common occurrences. Early attempts at signalling – a vital consideration of railway safety – were at best rudimentary, with signalmen positioned along the line holding flags or lamps. The introduction of the electric telegraph in 1848, the token system for single-line railways – which permitted only one driver to be in possession of the metal token issued for each stretch of track between passing stations to avoid head-on collisions – and John Saxby's invention of interlocking signals and point in 1856 were key principles in railway safety. The gradual introduction in the 1850s and 1860s of block signalling, which ensured that trains travelling in the same direction on double-tracked railway lines were kept far enough apart to avoid collisions, was a major advancement and became mandatory in 1889 following the Armagh disaster in Ireland when a runaway train killed seventy-eight passengers.

As we have read previously, Britain's worst railway disaster occurred during the First World War at Quintinshill, ten miles north of Carlisle, when correct signalling practice was not followed and the consequent death toll reached 215. Apart from this last accident the worst twentieth-century peacetime disaster before the 'Big Four' Grouping of 1923 was at Salisbury in 1906 when an express passenger train derailed at high speed and collided with another train, killing twenty-eight people. After the First Word War the incorrect use of a supposedly fail-safe single-line electric token brought

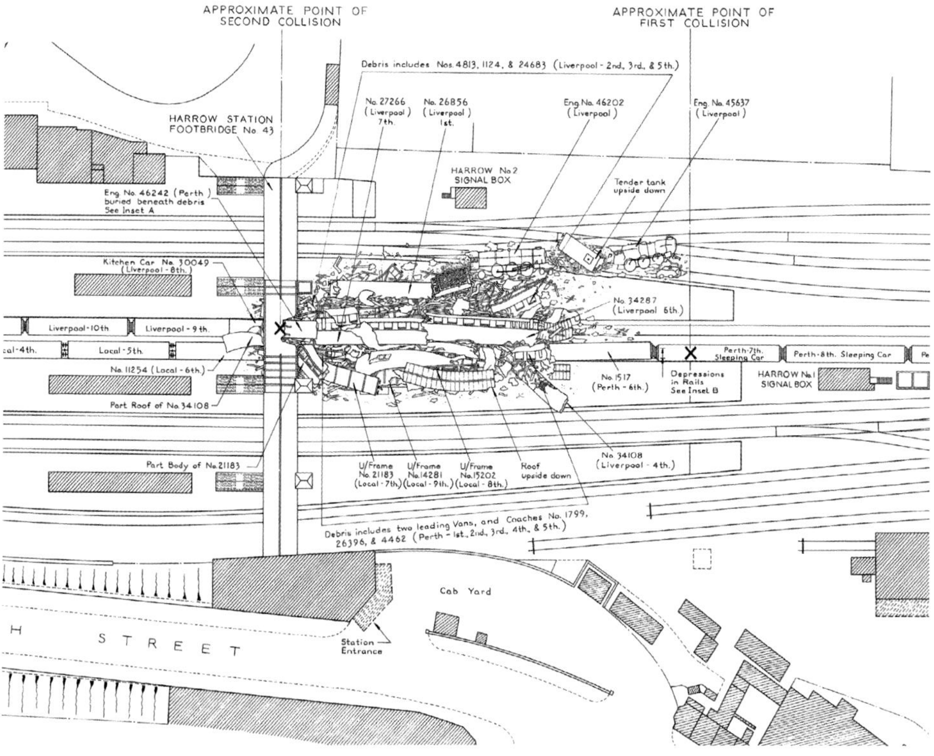

about a head-on collision between two trains at Abermule in mid-Wales in 1921, killing seventeen people.

During the era of the 'Big Four' (1923–1947) there were no less than fifty-seven serious railway accidents, of which twenty-two occurred during the years of the Second World War when railways were stretched to the limit and operated at night in blackout conditions. In total 478 people were killed with accidents at Darlington (1928/25 killed), Castlecary (1937/35 killed), Norton Fitzwarren (1940/27 killed), Eccles (1941/23 killed), Bourne End (1945/43 killed), South Croydon (1947/32 killed) and Goswick (1947/27 killed) being the worst incidents.

During the British Railways' era (1948–1994) there were 131 serious accidents in which a total of 631 people were killed. The first of these occurred at Winsford in Cheshire when a stationary train was run into by a following train.

Britain's worst peacetime railway disaster occurred at Harrow & Wealdstone station on 8 October 1952, when a stationary local train was hit in the rear by a southbound Perth to Euston sleeping car train. Seconds after the crash a northbound express hit the wreckage – 112 people were killed and 340 injured.

The second worst peacetime railway disaster happened in thick fog at St John's, Lewisham, on 4 December 1957 when a suburban electric train was hit in the rear by a steam-hauled Cannon Street to Ramsgate express – ninety people were killed and 173 injured. Driver error (the steam loco driver had passed a red signal), and the lack of an Automatic Warning System in the steam locomotive, were to blame.

Forty-nine people were killed at Hither Green in South London in 1967 when a train derailed at speed and thirty-five were killed at Clapham Junction in 1988 following a multiple collision between three trains. British Rail's final accident before privatization occurred at Cowden in Kent, when two diesel-electric multiple units collided head-on in thick fog on a single-line section – five people were killed.

LEFT: This official plan of the 1952 Harrow & Wealdstone accident shows the destruction caused when a stationary local train was hit by a southbound express which, in turn, was then hit by a northbound express – 112 people were killed.

BELOW: The remains of 'Battle of Britain' Class 4-6-2 No. 34083 *253 Squadron* after it had collided with the rear of an electric commuter train in thick fog at Lewisham on 4 December 1957 – ninety people were killed. The locomotive was repaired and remained operational until 1965.

1963 1964 1965 1966 1967 1968 1969 1970 1971 1972 1973 1974 1975 1976 1977

- **1963** – 'Beeching Report' published
- **1965** – British Rail double-arrow logo introduced
- **1966** – BR brand name *Inter-City* introduced
- **1966–1967** – Former Somerset & Dorset Joint Railway closes
- **1968** – The end of standard-gauge steam on British Rail
- **1971** – *King George V* becomes first steam locomotive to operate on main line since ban
- **1974** – Completion of West Coast Main Line electrifcation
- **1976** – Inter-City 125 trains introduced

DR BEECHING, CONTRACTION and THE END OF STEAM

1963–1993

1978	1981	1982	1983	1986	1987	1988	1989	1990	1991	1992	1993
First woman train driver on London Underground	Tilting Advanced Passenger Train (APT) introduced	BR regions replaced by business sectors	First woman train driver on British Rail	Tilting Advanced Passenger Train (APT) withdrawn		35 people killed in multiple train collision at Clapham Junction	Settle–Carlisle Line saved from closure		Last main line electric locomotive built in Britain	Last main line diesel-electric locomotive built in Britain	Railways Act ushers in privatization

The 'Beeching Report'

As we have read in the previous chapter, British Railways recorded its first operating loss in 1955, the same year of the announcement of the £1.24 billion 'Modernisation Plan'. In the face of outmoded working practices, overmanning and labour disputes, along with technology and infrastructure that harked back to the Victorian era, losses continued to mount. Labour relations were particularly bad and an ASLEF national rail strike from 29 May to 14 June 1955 did irreparable harm to the railways, especially on the freight side where already dwindling traffic was permanently lost to road transport – the strike was over pay differentials amounting to the price of a packet of cigarettes per week. Anthony Eden's new Conservative government caved in and BR had to pay the bill. BR's losses kept rising and by 1960 had reached £67.7 million per year or nearly £1.4 billion at today's prices.

Faced with this seemingly endless drain on the public purse Harold Macmillan's Conservative government grasped the nettle. The then Transport Minister, Ernest Marples, in that year formed the Stedeford Committee, an advisory group to report on the state of the railways and recommend workable solutions. One member of the committee was Dr Richard Beeching, the Technical Director of ICI, renowned for his analytical mind and undoubted business problem-solving skills. Beeching suggested that the railways should not be a public service but run as a profitable concern. While Beeching's ideas were not initially implemented Marples certainly took them on board and appointed him as chairman of the BR Board on 15 March 1961.

With Beeching now in the driving seat the outlook for Britain's railways was dire. Beeching commissioned a detailed survey of both freight and passenger traffic on BR within days of taking office – often misquoted, these surveys actually took place during the week of 17–23 April 1961 inclusive. In his eventual report, *The Reshaping of British Railways*, Beeching and his team recognized that 'it was impossible to perpetuate the massive recording effort involved'.

Almost two years then elapsed before the publication of his 'notorious' report on 27 March 1963. Priced one shilling and available from Her Majesty's Stationery Office, *The Reshaping of British Railways* was a collection of graphs, statistics, analysis and maps in two volumes that became commonly known as the 'Beeching Report'.

Part 1 comprised 148 pages of statistics and their analysis, effectively a death sentence for around 5,000 railway route miles and over a third of the country's stations. The report claimed that one-third of the route mileage carried only 1 per cent of total passenger miles and 1 per cent of freight ton miles on BR. With the 41-page Appendix 2 to the Report listing lines and stations throughout England, Scotland and Wales proposed for closure, the document was the death knell for Britain's rural railways.

Part 2 comprised thirteen large fold-out maps, produced to graphically illustrate the statistical information gathered for Part 1. Covering 'Density of Passenger Traffic', 'Density of Freight Traffic', 'Proposed Withdrawal of Passenger Services' and 'Proposed Modification of Passenger Train Services' it bore bad tidings for the railway industry. However, on the positive side, the map of 'Liner Train Routes and Terminals Under Consideration' was an important pointer to one of the

positive outcomes of the report – and the Freightliner train service is today the market leader for movement of short and deep-sea containers to and from England's ports. Another positive aspect was the widespread introduction of 'merry-go-round' trains between collieries and power stations.

The report's predominantly negative conclusions, dubbed the 'Beeching Axe' by the press, sparked a nationwide outcry, especially from rural communities who were most a risk. Despite this, Beeching's report claimed to provide all the answers, stating that 'most areas of the country are already served by a network of buses' and 'it appears that hardship will arise on only a very limited scale.' The consideration of hardship was the special responsibility of Transport Users Consultative Committees, where objections to closures could be lodged. If no bus service existed then, in some cases, replacement bus services had to be introduced. Ultimately it was for the Government – having also consulted its regional economic planning councils – to decide whether a line closed or not.

Fortunately a second Beeching report published on 16 February 1965, *The Development of the Major Railway Trunk Routes*, was not implemented by Harold Wilson's Labour Government – otherwise Britain's rail network would have been decimated. Beeching returned to a more lucrative career at ICI a few months later and was made a life peer. He died in 1985 aged 71.

The arch-villain of the piece was undoubtedly Ernest Marples, whose claim to fame previously as Postmaster-General had been to introduce Premium Bonds, STD telephone codes and postcodes. As Minister of Transport he had introduced parking meters, traffic wardens, the MOT test and yellow parking lines. However, there was an enormous conflict of interest when he was appointed to this post as he held 80 per cent of the shares of the motorway and construction company Marples Ridgway. He got over this little problem by selling them to his wife! In 1975 he hurriedly fled the UK to avoid expensive court cases and paying a large amount of capital gains and income tax. He spent his last years living at his vineyard in the Beaujolais wine region of France and died a rich man in 1978.

Farewell to the Former
Somerset and Dorset Joint Railway

Photo: I. Peters. *Block Courtesy: Railway Magazine.*

53807, the (then) sole survivor of the S.D.J.R. class 7F 2-8-0s at Masbury summit with the 8.55 a.m. down freight from Bath on the last day of its service, 5th September, 1962.

Photographic Souvenir

in connection with

LAST PASSENGER TRAIN

on the

Bath — Templecombe — Bournemouth Section

SUNDAY, 6th MARCH, 1966

Organised by the STEPHENSON LOCOMOTIVE SOCIETY
(Midland Area)

PREVIOUS SPREAD: Dai Woodham's scrapyard at Barry in South Wales started buying withdrawn locomotives from British Railways in 1959 and by 1968 had purchased 297. Of these, preservationists saved 213 although many spent years on 'death row'. Here, a row of former GWR small 'Prairie' tanks and 'Manors' stand rusting in the sea air in August 1979. The closest, 2-6-2T No. 5538, even has an original-style BR 'unicycling lion' emblem on its side tanks. Withdrawn in 1961, the locomotive was the 185th to be saved and is now being restored by the Dean Forest Railway.

FAR LEFT: Dr Richard Beeching proudly holds aloft his two-part *Reshaping of British Railways*, or the 'Beeching Report' as it is more widely known, on its publication day, 27 March 1963.

LEFT: One of the most famous victims of Dr Beeching's 'Axe' was the much-loved Somerset & Dorset Joint Railway which closed on 7 March 1966.

Mass railway closures

ABOVE: Seen here in 1971, Llangollen station was one of many that lost its passenger service on 18 January 1965 when the Ruabon to Barmouth line closed. Goods trains continued until April 1968 but since then the station, along with the line to Corwen, has been gradually reopened as a heritage railway.

RIGHT: Closure notices were pasted up at thousands of stations across Britain in the wake of the 'Beeching Report'. Here the line from Manchester Central to Stockport Tiviot Dale receives its death sentence that was eventually carried out on 2 January 1967.

Apart from some important exceptions the Beeching closure proposals went ahead, firstly under the incumbent Conservative Government but then, from 16 October 1964, under Harold Wilson's Labour Government who, despite their fierce criticism while in opposition, wielded the 'Beeching Axe' with gusto. Closures continued until the mid-1970s by which time over 4,000 route miles of railways had been closed along with around 2,500 stations, and nearly 68,000 railway jobs had been lost.

No region of Britain escaped but some fared better than others. In the Southwest of England thirty-seven lines closed including one, Coleford Junction to Okehampton, that had not been listed by Beeching. Fortunately for us today there were five survivors – the branch lines to St Ives, Looe, Gunnislake, Exmouth and Severn Beach were all reprieved due to the poor roads and the consequent difficulty of providing alternative bus services.

Of all the English regions the south, with its higher density of population and electrified lines, escaped mass closures but even so twenty-eight were closed including two that were not originally listed for closure. On a more positive note eight routes were reprieved including Ryde Pier Head to Shanklin on the Isle of Wight and Ashford to Hastings.

In Central England there was a blood bath with no fewer than forty-two lines being axed by 1970. Although not listed for closure the former GWR mainline south from Stratford-upon-Avon to Cheltenham and the cross-country Varsity Line between Oxford and Cambridge also bit the dust. By far the biggest casualty was the former Great Central Railway main line between Sheffield and Marylebone which, apart from a temporary reprieve for a short section, effectively closed as a through route in 1966.

On glancing at Map 9A of the 'Beeching Report' it would have appeared that Eastern England was going to avoid too much carnage but in reality the end result was very different. Of the twenty-two routes that had been closed by 1970, five were not on Beeching's original list –

these were concentrated in north Norfolk with three radiating out from King's Lynn. Although the East Lincolnshire line ceased to be a through route from 5 October 1970 the East Suffolk Line between Ipswich and Lowestoft was spared through ingenious management cost-cutting.

Over in Wales no fewer than twenty-seven lines had closed by 1970 leaving much of the principality devoid of services. The longest of these rural routes to disappear from the railway map of Britain were the three that served the county town of Brecon – the lines from Neath, Hereford and Moat Lane Junction had closed even before the publication of the 'Beeching Report', although they were all included in the closure proposals. Ruabon to Barmouth and Carmarthen to Aberystwyth were two other lengthy routes that disappeared from the timetables. Fortunately there were two notable survivors – the Conwy Valley Line from Llandudno Junction to Blaenau Ffestiniog and the Central Wales Line from Craven Arms to Llanelli were both reprieved due to poor local roads and the difficulty of running replacement bus services. Cynics might say that the Central Wales Line was only saved on two separate occasions because it passed through three marginal constituencies!

Although Dr Beeching's 'Axe' was wielded on thirty-nine routes in Northern England there were fortunately fifteen survivors. Notable among these were the Cumbrian Coast Line from Barrow-in-Furness to Whitehaven, the highly scenic Esk Valley Line between Middlesbrough and Whitby and last but not least the Settle–Carlisle Line, which escaped several closure threats before being reprieved by the government in 1989. One major route not listed for closure was the electrified Woodhead Line between Manchester and Sheffield, which was sacrificed in 1981 instead of the non-electrified Hope Valley route between the two cities.

Finally to Scotland, which suffered forty-four route closures including seven not originally listed by Beeching. While there was justifiable outrage at the closure of the Waverley Route between Edinburgh and Carlisle in early 1969, the same cannot be said about the 'Port Road' from Dumfries to Stranraer which sadly had already quietly slipped away in June 1965. Of course the former closure has had a 50 per cent happy ending, with the line reopening between Edinburgh and Tweedbank in 2015. There were seven lucky survivors from the 'Beeching Axe' in Scotland, all of which escaped closure due to strong public and political pressure. Notable among these are the long and meandering route from Inverness to Wick and Thurso, the highly scenic line from Inverness to Kyle of Lochalsh and the Ayr to Stranraer Harbour line. The latter's future does have a question mark over it south of Girvan as ferry services to Larne in Northern Ireland were moved from Stranraer Harbour to the non-rail connected port of Cairnryan in 2011.

The end of steam

At the time of the 'Modernisation Plan' of 1955 there were around 19,000 standard-gauge steam locomotives operating on British Railways and it was reckoned that diesel and electric locomotives would gradually replace steam haulage over the next thirty years. However, it soon became apparent that the changeover period was going to be much, much shorter. Apart from a few notable exceptions steam had virtually been eliminated on all mainlines around Britain by 1966. By then British Railways had become British Rail and so desperate was it to change its corporate image that the wholesale withdrawal of steam locomotives was high on the agenda, despite the fact that some of them were less than ten years old.

On the Western Region where diesel-hydraulics temporarily held sway, steam had been eliminated at the end of 1965 – the only pocket of resistance was on the 100 per cent steam-operated Somerset & Dorset route which held out until closure on 7 March 1966. However, the Southern Region continued with steam operations on the Waterloo to Bournemouth route until third-rail electrification had been completed in July 1967 – the sight and sound of Bulleid's Pacifics thundering through Basingstoke at the head of their passenger trains was the swansong of main-line steam in Britain.

Steam haulage continued on the Cambrian main line in Central Wales until December 1966 when Machynlleth shed closed. Croes Newydd (Wrexham) and Shrewsbury closed in June and November 1967 respectively.

In East Anglia steam had been eradicated as early as 1962 but in other parts of the Eastern Region it took four more years before they had gone for good. While King's Cross shed closed to steam in 1963, further up the line at Doncaster it clung on to life until May 1966. Further north in the North Eastern Region York shed lost its steam allocation in the summer of 1967 and Leeds Holbeck shed in October of that year. In the industrial northeast West Hartlepool, North Blyth, Sunderland and Tyne Dock

BELOW: Last days of steam on the Southern - looking remarkably clean but minus its nameplate (*Aberdeen Commonwealth*), 'Merchant Navy' Class 4-6-2 No. 35007 heads out of Waterloo station with the 8.35 a.m. train to Bournemouth on 28 June 1967.

RIGHT: The end of an era – suitably adorned to mark the end of steam on British Rail, Stanier '8F' 2-8-0 No. 48278 awaits its last duty at Rose Grove engine shed (10F) in Lancashire on 3 August 1968.

LAST STEAM
COPY PIT BANKER
THE END

sheds had already closed in September. The last two sheds to close in the North Eastern Region were at Royston in November and Normanton at the end of the year.

Steam also clung on to life in parts of Scotland with Nigel Gresley's 'A4' Pacifics performing their swansong at the head of the Glasgow to Aberdeen 3-hour expresses until September 1966. At the northern end of the West Coast Main Line, Glasgow Polmadie continued to supply steam locomotives until May 1967 as did nearby Corkerhill shed, Edinburgh St Margarets and all remaining Scottish steam sheds except for one. This dubious achievement went to Motherwell, which retained steam until the end of June.

By 1 January 1968 the end was nigh and there were only six steam sheds left operating in Britain, all of them on the London Midland Region in North West England. By 6 May three – Carlisle Kingmoor, Workington and Tebay – had closed leaving just Carnforth, Rose Grove and Lostock Hall still operational. Those last few months of steam drew railway enthusiasts in their thousands from all over Britain to witness the end of a glorious era. The end came on 11 August, the final day of steam haulage, and at midnight the last three sheds closed and steam had been eradicated for good – or so the enlightened management at British Rail thought!

To mark the end of steam on 11 August a special was run between Liverpool, Manchester and then over the Settle–Carlisle Line to Carlisle and return. Known as the '15 Guinea Special' it was hauled at various stages by three of Stanier's 'Black 5' 4-6-0s and 'Britannia' Class 4-6-2 No. 70013 *Oliver Cromwell*. Thousands of spectators lined the route thinking it was the last time they would ever witness main-line steam haulage in Britain. Nearly fifty years on how wrong they were!

Of course it mustn't be forgotten that BR continued to operate steam locomotives on its narrow-gauge Vale of Rheidol Railway in West Wales until 1989 when the line was privatized.

ABOVE: Giants at rest – during the final year of steam on British Rail former BR Crosti-boiler standard Class '9F' 2-10-0 92029 and Stanier '8F' 48626 await their next turn of duty at Birkenhead engine shed (8H) on 27 September 1967.

Barry Scrapyard

During the 1960s scrapyards up and down the country worked non-stop to break up and dispose of withdrawn steam locomotives. However, there was one scrapyard in South Wales where time stood still and several hundred rusting hulks waited patiently to be saved from death row.

Dai Woodham's scrapyard at Barry took its first delivery of withdrawn Western Region steam locomotives as early as 1959 but initially the numbers were low so scrapping was able to keep pace. As time went on the numbers steadily increased and by 1968 Woodham's had bought 297 withdrawn steam locomotives, not only from the Western Region but also the London Midland and Southern regions, and his site at Barry was filled to bursting. The reason for this was that Woodham had also bought large numbers of redundant goods wagons from BR and scrapping these was much easier than breaking up steam locomotives so the latter were put on an extended death row to await their turn. However, by the late 1960s the steam preservation movement was gathering pace and eventually 213 of Woodham's locomotives, many of them by now rusting hulks, were sold for restoration – the oldest inhabitant at Barry arrived in 1961 and was finally saved in 1986, twenty-five years later. Today's heritage railways and mainline steam charter organizations have much to thank the Welshman for.

RIGHT: A request that could not be ignored – even in 1981 there were still a good number of withdrawn steam locomotives waiting to be rescued from Dai Woodham's scrapyard in South Wales.

BELOW: Bulleid 'Pacifics' and BR Standard locomotives lie rusting away at Barry in August 1979. In the foreground is 'Merchant Navy' class 4-6-2 35025 *Brocklebank Line*. After twenty-one years at the scrapyard this locomotive was saved at the eleventh hour from scrapping and is now being restored in Kent.

A new identity

With steam on the way out, British Railways made an effort to change its outdated public image and in 1964 commissioned the Design Research Unit to overhaul its house style and logo. In addition to renaming itself British Rail the end result of BR's design consultation, which was launched at the beginning of 1965, was a new typeface ('Rail Alphabet'), the double-arrow logo (soon nicknamed the 'arrow of indecision') and new corporate colours. So wide-ranging was this new face of BR that a 4-volume corporate identity manual was issued in instalments between July 1965 and April 1970 – nothing was overlooked! While BR and its new identity died in 1994, the new double-arrow logo has stood the test of time and fifty years later is still widely used on signage across Britain.

Preceded by the XP64 experimental corporate livery, the standardized 'Rail Blue' livery was introduced in 1965 while the 'Inter-City' brand name was first introduced in 1966 for its long-distance expresses – a new livery of two-tone blue and pale grey was applied to coaches on these routes with 'Inter-City' picked out in white.

The introduction in 1976 of Inter-City 125 High Speed Train (HST) sets on the Western Region main line from Paddington to Bristol, Cardiff and Swansea was a great success. With a maximum speed of 125 mph (compared to earlier diesels' 100-mph limit) and vastly improved passenger comfort, soon passengers were attracted back to Britain's much-maligned railways – helped along the way by vigorous marketing to 'let the train take the strain'. The HSTs were also operating within a few years on the East Coast Main Line from King's Cross, to Edinburgh, Aberdeen and Inverness; the Midland main line out of St Pancras; as far west as Swansea and Carmarthen; and to the south western tip of England at Penzance. Forty years on these highly successful trains are still in service between Paddington and Bristol, Cardiff, Swansea and the West Country. The completion in 1974 of West Coast Main Line electrification also saw vastly improved services on this important Anglo-Scottish railway artery.

The successful rebranding of Britain's inter-city network was closely observed by other countries, with the brand name being launched in West Germany in 1971 to replace the former Trans Europe Express trains – Deutsche Bundesbahn even paid a royalty for using the name to British Rail for many years. Intercity-Express (ICE) high-speed trains introduced in 1991 now operate throughout Germany while the EuroCity name brand is used by international train services within Europe.

LEFT: Introduced in 1965, the new modern house style for British Rail included the famous double arrow logo, dubbed the 'arrow of indecision', which is still widely used on signage for railway stations.

BELOW: The introduction of the Inter-City 125 High Speed Train in 1976 was a great success for British Rail, with an increasing number of passengers wooed over to the new service by a high-powered publicity campaign.

Advanced passenger train

By the early 1970s the new Japanese Shinkansen high-speed trains were having a major impact on railway engineers around the world. In France SNCF were working on their TGV prototypes and the age of high-speed rail travel was becoming a reality. Designed to run on the existing West Coast Main Line, the Advanced Passenger Train (APT) was Britain's response to this quest for speed and by 1972 the prototype gas turbine four-car APT-E was ready for testing. The train was fitted with a tilting mechanism that allowed it to travel up to 155 mph on existing curving track and in tests over the following four years this new technology worked well. The next stage in the development of the APT was the building of three 14-car electric train sets designated APT-P but these had two power cars located in the middle of the train, effectively cutting the train in half with six articulated passenger coaches on either side.

The APT-P was launched between Glasgow Central and London Euston in December 1981 but failures with the tilting mechanism and brakes saw the service suspended after only a few days. The British press had a field day dubbing it the '**A**ccident **P**rone **T**rain'. With the problems corrected the train was reintroduced in 1984 and operated successfully but by then interest in the project had waned and the proposed fleet of APT-S production trains was cancelled. The APT-P trains continued in service until 1986 and then were quietly withdrawn. Despite this expensive failure the design elements for the later BR Mk 4 coaches and Class '91' electric locomotives now used on the East Coast Main Line are direct descendants of the APT project. In the end it was an Italian company that successfully developed the tilting train technology – introduced in 2004, the Class '390' 'Pendolino' now in service on the West Coast Main Line are made by French-owned Fiat Ferroviaria.

BELOW: Fitted with a tilting mechanism to allow higher running speeds on the existing rail network, the new Advanced Passenger Train gas turbine prototype is seen here on the right alongside the prototype HST 125.

Marshalling yards and freight services

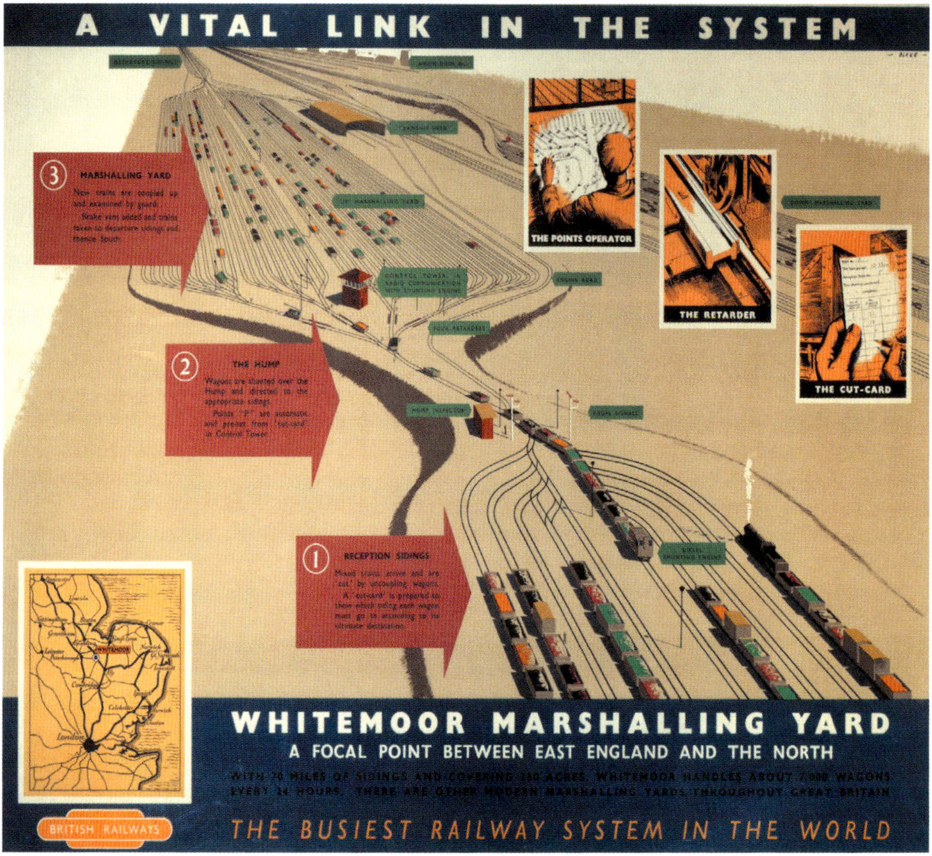

Moving vast amounts of freight in separate wagonloads was the norm on Britain's railways until the 1960s. To handle and sort these wagons into specific trains was a time-consuming process and even before the 'Big Four' Grouping of 1923 several railway companies had built marshalling yards at strategic locations on their network. The Great Central Railway opened the enormous hump yard at Wath in South Yorkshire in 1907 to handle coal from no fewer than forty-five collieries operating within a 10-mile radius – its thirty-six miles of track contained two yards, each with thirty-one departure sidings, and could handle up to 5,000 wagons each day. Opened in 1917 the London & South Western Railway's hump marshalling yard at Feltham in southwest London had two gravity shunts, electrically operated points, contained thirty-two miles of track and could handle over 3,000 wagons each day. By the 1950s the former Midland Railway's marshalling yard at Toton in Nottinghamshire handled over one million wagons per year and was the largest of its kind in Europe.

In Britain marshalling yards were operated using the 'hump' method. Here incoming trains arrived in reception sidings where wagons would be uncoupled and then shunted forward over a hump before running by gravity into marshalling sidings to await their journey to their next destination. Each wagon would have a card clipped to it giving details of its origination and destination so that whole trains could be made up of wagons with the same destination. It was not only a complicated business but also very dangerous as each moving wagon had to be

slowed by pinning down the brakes and directed into its correct siding by men running alongside it. The use of automatic hydraulic wagon retarders was first introduced by the LNER at its new Whitemoor Yard in 1930, and electrically operated points saved numerous lives among the shunting fraternity although this modernzsation was not completed until the 1960s – Whitemoor Yard at March in Cambridgeshire was once the largest in Britain and the second largest in Europe.

As we have read in the previous chapter, the 'Modernisation Plan' of 1955 put into place the resiting and modernizing of marshalling yards. At that time there were often two or more yards, of old-fashioned design, situated close together and many of them were of the flat type. A total of fifty-five new hump yards were planned with power-operated points and mechanical wagon retarders controlled from one central position – the cost for all of this in 1955 was reckoned to be £80 million. Despite the continuing loss of freight traffic to road transport the plan went ahead and by the 1960s a good proportion of new marshalling yards had been built and were operational – notable among those that opened in 1963 were Carlisle Kingmoor, Millerhill in Edinburgh, Healey Mills in Wakefield, Tees Yard in Middlesbrough and Tyne Yard in Newcastle. Opened by Dr Richard Beeching in 1965, the hump yard at Sheffield Tinsley contained fifty sorting sidings and all wagon movements were controlled by a new computerized control system.

Sadly, by the time these vast new yards had opened the freight that they were built to handle was already dwindling away and within ten years they had become obsolete. In some ways the 1963 'Beeching Report' made matters a lot worse, virtually doing away with many of the goals of the 'Modernisation Plan' and closing around 4,000 route miles of railway, thereby cutting off the branches that fed the main trunk routes. By 1963 there were still around 850,000 individual wagons on BR but wagonload freight was now virtually dead in the water, with more flexible road transport taking over this once-important side of the business. A start had already been made to win back traffic with the introduction of the 'Condor' express container train between London and Glasgow in 1959 and a second route, from Birmingham to Glasgow, was added in 1963.

In other ways the 'Beeching Report' correctly foresaw the primary freight role of the railways as handling trainloads of bulk freight such as coal and mineral traffic, 146 million tons (in 1961) and fifty-four million tons (in 1963) respectively, recommending the introduction of 'merry-go-round' trains for coal traffic between collieries (there were 620 in 1961) and power stations. However, the most important proposal from Beeching was for the introduction of intermodal liner trains carrying intermodal shipping containers, which his Report correctly forecast could see a transfer of trunk haul freight traffic from road and on to rail. Taking over from the pioneering 'Condor', the first Freightliner service started in 1965 and fifty years on at least two of Dr Beeching's recommendations have proved highly successful, with increasing amounts of Britain's long-distance freight traffic now carried by rail.

FAR LEFT: A poster produced by British Railways in the 1950s to promote the efficiency of the Whitemoor Marshalling Yard at March in Cambridgeshire. The poster shows the layout of the yard and details on the reception sidings, the Hump, wagon retarders and the marshalling yard sidings.

ABOVE: Terence Cuneo's poster produced in 1960, 'Night Freight', depicts the new 'Condor' overnight express container service between London and Glasgow hauled by a pair of the new 'Metrovick' Co-Bo diesel locomotives.

Railway preservation

In what was a truly British phenomenon, railway preservation had its beginnings in 1951 when a group of enthusiasts led by the writer Tom Rolt saved the narrow-gauge Talyllyn Railway in West Wales from closure, becoming the first railway preservation scheme in the world. With railway closures gathering pace across Britain, the Bluebell Railway in Sussex became the first preserved standard-gauge steam-operated passenger railway in the world to operate a public service when it opened in 1960. Both of these pioneering railways are flourishing today.

Back in North Wales the narrow-gauge Ffestiniog Railway had closed in 1946 but a group of enthusiasts led by Alan Pegler raised enough money to pay off its outstanding debts and in 1955 were able to reopen a short section of the line between Porthmadog and Boston Lodge. As restoration and track laying progressed the railway continued to expand, reaching Tan-y-bwlch in 1958 and Dduallt ten years later. From here a new 2½-mile deviation, including a spiral and tunnel, was built taking the little railway above a new hydro-electric scheme. Tanygrisiau was reached in 1978 and the railway finally reopened to Blaenau Ffestiniog in 1983. Now a major tourist attraction, the Ffestiniog Railway's rebirth was nearly all achieved by a dedicated band of unpaid volunteers.

From these early beginnings over fifty years ago, the preservation movement has grown from strength to strength, with currently around 130 heritage railways and museums in Britain. Many of the reopened lines that had been victims of Dr Beeching's 'Axe' back in the 1960s have been brought back to life by dedicated enthusiasts – all tourist attractions in themselves and major contributors to their local economies, the Severn Valley Railway, North Yorkshire Moors Railway, West Somerset Railway, North Norfolk Railway and Bodmin & Wenford Railway are just a few examples of this post-Beeching renaissance.

While some of Britain's preserved railways remain isolated from the national rail network a lucky few benefit greatly from their physical connections, not only allowing the through working of charter trains from other parts of Britain but also ease of access for visitors. The Mid-Hants Railway, North Yorkshire Moors Railway, North Norfolk Railway and Mid-Norfolk Railway all benefit from this, while the recent connection of the Bluebell Railway to the national network at East Grinstead has seen rising visitor numbers. In the very near future visitors to the Swanage Railway will once again be able to travel down from Waterloo by changing trains at Wareham station.

The railway preservation movement was also greatly assisted by a Welsh scrap merchant, Dai Woodham, who, during the 1960s, bought a total of 297 withdrawn steam locomotives from British Railways, of which 213 were subsequently sold to preservationists for use on the growing numbers of preserved lines across the country. After years of painstaking restoration, many of these steam locomotives can now be seen at work not only on heritage railways but also on a growing number of mainline charter operations. The culmination of the steam preservation movement in Britain came in 2008 when the brand-new 'A1' Class 4-6-2 No. 60163 *Tornado* emerged from Darlington Locomotive Works. A product of eighteen years' work at a cost of £3 million, this superb machine has not only put in many fine performances on the main line but also draws enormous crowds wherever it appears. At the other end of the scale GWR steam railmotor No. 93 was restored to working order with many new parts and since 2012 has been in action in Devon and Cornwall.

Since the success of *Tornado* there are now many 'new-build' steam locomotives currently under construction or at planning stage – these include a GWR 'Saint', GWR 'Grange', GWR 'County', LMS 'Patriot' and of course the mighty LNER 'P2' 2-8-2.

The story of British railway preservation schemes culminated in the rebuilding of the long-closed Welsh Highland Railway in North Wales. The 25-mile narrow-gauge line was reopened in stages between 1997 and 2011 and now gives passengers a highly scenic 2½-hour journey through the Snowdonia National Park between Caernarfon and Porthmadog. Massive (by narrow-gauge standards) British-built articulated steam locomotives repatriated from South Africa haul trains on this winding and steeply graded route.

RIGHT: Most visitors to the Isle of Wight go for the warm summer temperatures, beautiful sandy beaches and unspoilt countryside, but a mid-November morning like this takes some beating. Adams class 'O2' 0-4-4T No. 24 *Calbourne* looks very much as if it is heading an early morning Ryde Pier Head to Newport and Cowes service in the 1960s, but this in fact was a photographic charter on 16 November 2010. The train is seen emerging from the swirling morning mist near Ashey station on the Isle of Wight Steam Railway.

The death of British Rail

In a move that was widely seen as a prelude to the privatization of the railways, the regions of British Rail were replaced by business sectors in 1982. On the passenger side InterCity (now without the hyphen) was already in control of express services, while the new Network SouthEast (replacing the short-lived London & South Eastern sector in 1986) operated London commuter services (and further afield) and the highly subsidized Regional Railways took control of all remaining regional services.

With its own distinctive livery of red, white and blue the Network SouthEast sector covered a wide area from the Midlands to Eastern England, from the Thames Valley to the South Coast and as far west as Exeter via Salisbury. A major replacement programme for the ageing rolling stock, coupled with further third-rail electrification and the introduction of the popular Network Railcard under the business-savvy BR Sector Director Chris Green, all contributed to its later success.

The Regional Railways sector was the Cinderella of British Rail, having to make do with second-hand locomotives and rolling stock or worn-out first-generation diesel multiple units on its vast sprawling network that stretched from Scotland, northern England and the Midlands to East Anglia and Wales. The rolling stock issue was slowly dealt with by the introduction of new bus-type four-wheel Pacers between 1984 and 1987 and the more comfortable Sprinter diesel multiple units between 1987 and 1989. While both these types remain in operation today, the Pacer's unsuitability for modern operations has even been raised in the House of Commons! By its geographical nature Regional Railways was never a profit-making concern, depending on Government subsidies of up to 400 per cent of its annual revenue to keep trains running on rural railways – it is a reassuring fact that there were no rail closures following the creation of Regional Railways despite some obvious loss-making candidates such as the Central Wales Line and routes in the Far North of Scotland.

On the freight side the Railfreight sector was also created in 1982 and a new corporate identity was introduced, with locomotives and rolling stock being

finished in 'Railfreight Grey' and a stripe along the lower sides of some of the former picked out in 'Railfreight Red Stripe'. Railfreight was split into four sectors in 1987: Trainload Freight handled bulk trainloads of coal, aggregates and metal; Railfreight Distribution handled non-trainload freight; Freightliner was in control of intermodal container traffic; and Rail Express Systems handled parcels traffic. A new two-tone 'Railfreight Grey' colour scheme was also applied to locomotives.

After four years of Thatcher Government Britain's railways were in a poor shape with annual losses approaching £1 billion, passenger numbers declining and labour disputes. Billed by some as the 'Second Beeching Report', a report into the state of British Rail and its future viability commissioned by the Conservative government was published in 1983 – the Serpell Report put forward various options to reduce BR's deficit, with the worst case scenario being the closure of over 80 per cent of Britain's railways leaving just three main routes radiating out from London, namely the West Coast Main Line, the East Coast Main Line as far as Newcastle and the former GWR route to Bristol and Cardiff. Other less stringent options would have still seen wholesale rail closures and even the government baulked at those. Fortunately Serpell was quietly forgotten and over the next decade British Rail's fortunes slowly began to improve thanks to investment in the infrastructure and new rolling stock.

Of note during this period was the electrification of the East Coast Main Line between London King's Cross, Leeds and Edinburgh, which was completed in 1991 after a 6-year construction period. Network SouthEast benefitted greatly, with third-rail electrification extended to Hastings and Weymouth and new rolling stock replacing the old slam-door coaches that had been around for decades. Once threatened with closure Marylebone station in London was given a new lease of life in 1987 when an upgrading of the Chiltern Main Line via High Wycombe, Bicester and Banbury allowed the introduction of services to Birmingham. Electrification of the Midland Main Line between St Pancras and Bedford, the reopening of Snow Hill Tunnel and the introduction of dual voltage trains opened up the 140-mile cross-London Thameslink route in 1988, allowing passenger trains to run between Bedford and Brighton.

By the early 1990s British Rail was on the road to a brighter future but this final chapter in the history of Britain's nationalized railways effectively came to a close on 5 November 1993 when the Railways Act was passed, ushering in a new era of privatization.

LEFT: The last decade of BR – observed by rail fans at the platform end, No. 47571 departs from London Liverpool Street with a Norwich service while No. 47569 stands on the stabling point, 20 April 1985.

BELOW: The final year of British Rail – change is afoot for Skipton station in West Yorkshire in April 1994. Regional Railways class 156 'Sprinter' 156 487 and a 'Pacer' unit are forming a local service to Leeds. The station still retained the Yorkshire flagstones on some of the platforms and full resignalling and electrification beckoned in the months ahead under the ownership of Railtrack PLC.

1994

Channel Tunnel opens

First stages of privatization enacted

1999

31 people killed in Ladbroke Grove train collision

2000

Four people killed in Hatfield train accident

2002

Network Rail takes over from bankrupt Railtrack

2003

Phase 1 of HS1 opens

2004

'Pendolino' tilting trains introduced

Last Travelling Post Office train

2007

Phase 2 of HS1 opens

PRIVATIZATION and HIGH-SPEED

1994 onwards

2009	2014	2015	2017	2019	2020	2026	2031
New-build steam locomotive 'Tornado' enters service	Edinburgh Trams opens	Borders Railway reopens between Edinburgh and Tweedbank	Great Western electrification due to be completed	Crossrail due to be completed	Midland and Northern Hub electrification due to be completed	Phase 1 of HS2 due to be completed	Phase 2 of HS2 due to be completed

The Channel Tunnel

As early as the beginning of the nineteenth century there were plans to build a tunnel under the English Channel, however, strained relationships between Britain and France halted these proposals. A number of other proposals during the course of the century also failed to come to fruition despite initial geological surveys having been undertaken. The enterprising chairman of the Manchester, Sheffield & Lincolnshire Railway, Sir William Watkin, later in the century, planned a grand continental railway linking northern England with mainland Europe via a Channel Tunnel, but the project was abandoned in 1882 despite the exploratory work on both sides of the Channel. Watkins' Great Central 'London Extension', which opened in 1899 from the north to Marylebone, was still built to the continental loading gauge, however.

Further proposals by the British government for a Channel Tunnel were abandoned after the First World War, as was a later proposal in 1929. Construction of a UK-France government-backed scheme was even started in 1974 but was cancelled (by the British Labour Government) a year later, shortly after tunnelling operations had begun.

A privately funded scheme was finally given the green light at the Treaty of Canterbury in 1986. Project responsibility was split between British Channel Tunnel Group (two British banks and five construction companies) and the French Tranche-Manche (three French banks and five construction companies), and the £2.6 billion raised by private finance was unprecedented for such a project at that time. Tunnelling began in 1988, being completed in 1994 with a cost over-run of 80 per cent. There were ten fatal accidents during its construction out of a workforce of 15,000. The 31-mile tunnel comprises three separate parallel bores; two outer bores for railway lines and a smaller inner one for service and emergency vehicles. Eurotunnel Shuttle trains convey road vehicles between terminals at Cheriton, near Folkestone, and Coquelles, near Calais; the Eurostar services between Paris, Brussels and London; and through international freight services. Plans for Deutsche Bahn to operate an Intercity-Express (ICE) service between Frankfurt, Amsterdam and London are likely to take shape in the near future and through trains between London, the South of France and Spain are also another exciting possibility.

When the tunnel opened in 1994 there was no high-speed rail link between Folkestone and London, although in France the Eurostar trains were able to use the newly built LGV Nord line to Paris. For seven years Eurostar trains were forced to travel at much slower speeds on the existing third-rail network between Folkestone and a temporary terminal in London at Waterloo International. It took until 1996 for construction of the high-speed rail link (HS1) between Folkestone and London St Pancras International to get underway and its completion in 2007 gave Britain, rather belatedly, its first high-speed rail line.

Eurostar services are currently operated by triple-voltage 18-car Class 373 trains with a top speed of 186 mph although a speed limit of 99 mph exists in the tunnel. New Velaro e320 (Class 374) dual-voltage 16-car train sets built by Siemens with a top speed of 199 mph are expected to enter Eurostar service at the end of 2015.

PREVIOUS SPREAD: As the mist rises from the River Eden on 17 January 2004, a diverted West Coast Main Line Anglo-Scottish Virgin 'Voyager' service heads south through the Eden Valley at Armathwaite on the Settle–Carlisle line.

BELOW: The entrances to the Channel Tunnel on the English side during construction in 1991. The two outer tunnels are for southbound and northbound trains respectively while the smaller central tunnel is for service and emergency vehicles.

High Speed 1

While the French had already opened their high-speed rail link between the Channel Tunnel and Paris in 1994 and the Belgians a similar link to Brussels in 1997, the British lagged behind, forcing Eurostar trains to travel between the Channel tunnel and London on the existing third-rail network which limited speeds to 100 mph. The situation was far from ideal and it took until 1996 before a new high-speed line (HS1) was authorized by Parliament. Costing £80 million per mile the 67-mile electrified line was built in two stages, with the first section across North Kent tunnelling through the North Downs in a 2-mile tunnel and crossing the River Medway on a ¾-mile-long bridge. Serving Ashford International station and with a maximum line speed of 186 mph this opened in 2003, although Eurostar trains still needed to use the existing third-rail network to reach Waterloo International.

Phase 2 of HS1 was a major engineering feat involving the construction of three viaducts and a further 13½ miles of tunnels under the River Thames and East London in order to reach St Pancras station. Here, a new terminus was built in the grand Victorian station that had been saved from demolition in the 1960s by preservationists led by John Betjeman, subsequently to become Poet Laureate. With intermediate stations at Ebbsfleet International and Stratford International and a maximum line speed of 143 mph, Phase 2 opened in 2007, cutting the journey time between London and Paris to 2 hours 15 minutes. Domestic services using Class 395 Javelin trains between St Pancras and Dover/Ramsgate operated by TOC Southeastern also use HS1 for part of their journey.

Built with long passing loops that allow faster passenger trains to overtake, HS1 is also used by international freight services including through intermodal trains between continental Europe and the UK. Here, international terminals are located at Dollands Moor in Folkestone and Ripple Lane in Barking, while connections in London allow freight trains to use the West Coast Main Line, the North London Line and the Great Eastern Main Line. Maintenance of Eurostar trains is carried out at Temple Mills near Stratford International.

BELOW: Two of the 18-car Class '373' trains operated by Eurostar on their services to Paris and Brussels at rest inside St Pancras International station.

Privatization of Britain's railways

Margaret Thatcher's policies of the 1980s were influential in the lead up to the privatization of Britain's railways, but it was not until John Major came into office and oversaw the passing of the 'Railways Act' that the state industry of some 45-years' standing was sold off piecemeal. As a precursor to this, some minor business units had already been disposed of such as Sealink, British Transport Hotels and the railway works of British Rail Engineering Ltd. The restructuring of freight and passenger railway operations into business units such as Railfreight and InterCity also smoothed the path for the eventual transition. The Labour party had consistently opposed the privatization and had been vehement in their opposition to it, indeed previously objecting to the brutal rationalization of the railways under Dr. Beeching, but their term of office between 1997 and 2010 saw no change to the status quo.

The Office of the Rail Regulator was set up to regulate the entire railway industry following the controversial break-up of British Rail into more than 100 individual companies. The awarding of the twenty-five passenger train franchises was overseen by the Director of Passenger Rail Franchising but this did not include infrastructure such as stations, signalling and trackwork, which was sold to Railtrack, which went on to lease all but major stations to the newly franchised passenger train operators. The maintenance and renewal of the infrastructure was also let out to a multitude of small infrastructure companies formed on the back of privatisation. Just two Freight Operating Companies (FOCs) emerged from the sale of Transrail, Mainline, Loadhaul, Rail Express Systems, Railfreight Distribution and Freightliner freight business units, namely English Welsh & Scottish Railway and Freightliner. The varied fleet of British Rail locomotives, multiple units and passenger coaches was distributed between three Rolling Stock Leasing Companies (ROSCOs).

The consequent twenty-five Train Operating Companies, which in fact owned nothing, ran scheduled passenger services on approved routes. They paid Railtrack for the use of its infrastructure, hired-in their locomotives and rolling stock from ROSCOs, and sub-contracted all train maintenance and on-board catering.

BELOW: A Glasgow Central to London Euston Virgin 'Pendolino' speeds across the River Esk on the West Coast Main Line at Floriston near Gretna on 22 December 2010.

The Rail privatization of 1 April 1994 was slow to develop and took some time to bear fruit. Indeed, the new Labour Government saw the Transport Act 2000 enacted whereby a non-departmental public body, the Strategic Rail Authority (SRA), was established to guide the privatized rail industry. However, the passing of the Railways Act 2005 saw the SRA wound up in December 2006, its duties being taken over by the government's Department for Transport, Network Rail (the successor to Railtrack) and the Office of Rail Regulation, with limited powers also being devolved to the then Scottish Executive, the Welsh Assembly and the Greater London Authority.

Controversy abounds over railway privatization and whether the service has actually improved, given that the degree of customer service and quality of on-board passenger facilities can be called into question in some areas of the country. Standard 'walk-on' rail fares are considered to be amongst the highest in Europe, although highly competitive advance fares are eagerly pursued using a number of competitive internet-based company websites. Although there has been a marked increase in passenger footfall on rail services, arguably this may well be attributed to the period of good economic development and ever-increasing congestion on Britain's roads.

On the freight side, actual freight tonne miles conveyed has shown a considerable increase following the more straightforward manner in which rail freight was privatized. The closure of Britain's deep coal mines and the consequent increase in rail movements of imported coal over literally hundreds of miles has had its part to play in rail-freight revenues, but certainly a buoyant economy has influenced the growth.

In 1994, the last complete year of public ownership, the annual cost of British Rail to the taxpayer was £1.6 billion but the taxpayer today still subsidizes the country's 'privatized' railways to the tune of around £4 billion.

Community railways

Set up in 2005, the Association of Community Rail Partnerships is funded by the Department of Transport and today supports over thirty local rail routes around England and Wales. The majority are community railway lines, such as the Tarka Line to Barnstaple in Devon, but where more than one TOC runs services, such as on the Poacher Line to Skegness in Lincolnshire, these lines have community rail services.

Community rail partnerships offer a bridge between the railway and local communities, bringing together a wide range of interests along the rail corridor. Their work includes improving bus links to stations, developing walking and cycling routes and real-ale trails, brightening up station buildings and organizing art and education projects. Some rail partnerships have been instrumental in bringing about spectacular increases in rail use through innovative marketing and improved station facilities – for example, the Looe Valley Line in Cornwall (once threatened with closure by Dr Beeching) has seen passenger numbers increase by 50 per cent in the last ten years.

The scenic Looe Valley Line in Cornwall was once threatened with closure but in more recent times has experienced an upsurge in passenger numbers.

Railtrack, Network Rail & rail safety

Founded in the wake of the Railways Act 1993 when Britain's railways were privatized, Railtrack took control of all railway infrastructure including track, signalling, tunnels, bridges and most of the stations on 1 April 1994. Two years later the company was floated on the Stock Exchange. Income was derived mainly from payments that were made by the Train Operating Companies for operating trains on the company's tracks. Although safety was of paramount importance, a series of fatal rail crashes over the following six years brought Railtrack to its knees.

Seven people were killed and 139 injured on 19 September 1997 at Southall on the Great Western Main Line when an InterCity 125 travelling at 125 mph slammed into a freight train that was crossing its path. The driver of the express had passed two orange cautionary signals and one red danger signal due to a faulty on-board Automatic Warning System (AWS). In this case Railtrack was not to blame but the train operator, First Great Western, was fined £1.5 million.

A much worse accident happened on 5 October 1999 at Ladbroke Grove on the approaches to London's Paddington station when a westbound Thames Trains diesel multiple unit passed a poorly positioned red danger signal and collided head-on with an incoming InterCity 125. Spilled diesel fuel ignited and in the inferno thirty-one people were killed and 523 injured. At a subsequent public inquiry, blame for his horrific crash was shared between Thames Trains for insufficient driver training and Railtrack for the poorly positioned signal – the latter was severely criticized for its maintenance of the railway infrastructure and its safety record. Its successor, Network Rail, had to pay a fine of £4 million and Thames Trains was fined £2 million. As a result of the inquiry the Rail Safety & Standards Board and the Rail Accident Investigation Branch were subsequently created.

Just over a year after the Ladbroke Grove crash there occurred another serious railway accident, which sealed the fate of Railtrack. On 17 October 2000 a northbound GNER InterCity 125 train travelling at over 110 mph derailed on its approach to Hatfield station. Four people were killed and over seventy injured, and in the subsequent inquiry it was found that a fractured rail had caused the derailment and that the deaths had all occurred in the restaurant car, which after derailing had collided with an

overhead gantry. At a court case held in 2005 Network Rail (the successor to Railtrack) was found guilty of breaking health and safety regulations and its sub-contractor at the time of the accident, Balfour Beatty, pleaded guilty to the same charges.

The implications of the Hatfield accident were enormous as thousands of miles of track around Britain were required to be checked for metal fatigue, causing widespread disruption for over a year. With more than 1,000 speed restrictions in place, trains delayed and cancelled, and a costly track-replacement programme under way it was reckoned to be costing the UK economy over £40 million per week.

The relationship between Railtrack and the Rail Regulator had already been disintegrating but the Hatfield accident and its aftermath was the final straw for the company. Track renewals and compensation for the accident, coupled with out-of-control costs for rebuilding the West Coast Main Line, led the company to ask the Government for financial help. Part of the subsequent taxpayer bail out was then used to pay a £137 million dividend to its shareholders! Not surprisingly Railtrack was declared bankrupt in 2001, and while it was in administration the Government was forced to plough in a further £3.5 billion in order to keep the railways running. The end of this particular story came in October 2002 when the newly formed 'not for profit' government-created company of Network Rail bought Railtrack's assets.

Network Rail (NR) is based at The Quadrant:MK in Milton Keynes and today has a workforce of 34,000 and a revenue from the passenger and freight-train operating companies of £6.2 billion. It owns around 20,000 miles of track, 40,000 bridges and tunnels and over 2,500 stations. The vast majority of the latter are managed by train-operating companies although NR still operates most of the larger and busiest stations in London and major cities. NR is responsible for signalling and operates all the telecommunications needed to run the railway network, making it the largest private telecoms operator in Britain. It also owns a large fleet of rolling stock and a collection of ageing diesel locomotives, electro-diesels and diesel and electric multiple units that are used around the network for safety checks. Most of these are painted yellow, including Class '31' and Class '37' diesels and a converted HST 125 set. Although it is a not-for-profit company limited by guarantee, Network Rail was reclassified in 2014 as a central government body, consequently adding over £30 billion to the UK's public sector debt.

LEFT: A historic moment – the first ballast drop by Railtrack on the first newly laid section of HS1 near Ashford utilizing HQA Autoballaster wagons, 30 November 2001.

ABOVE: A historic moment on the Borders Railway – Direct Rail Services Class 37 diesel No. 37604 at the head of the 17.00 Tweedbank to Millerhill track testing train near Bowshank Tunnel on 12 May 2015.

Crossrail

Proposals for east-west and north-south underground railways with large-diameter tunnels under London were first put forward during the latter years of the Second World War. However, it was not until 1988 that the north-south Thameslink route opened, linking Bedford in the north with Brighton in the south via the reopened Snow Hill tunnel that runs beneath Smithfield meat market near the City of London.

Designed to take pressure off the existing overcrowded London Underground lines, the east-west route connecting Paddington and Liverpool Street stations had a much longer gestation period. While a 1989 proposal estimated such a link would cost £885 million, a private bill promoted by British Rail and London Underground and submitted to Parliament in 1991 was rejected. A further proposal put forward in 2001 also fell by the wayside but finally the Crossrail Bill was put before Parliament between 2005 and 2007, receiving Royal Assent on 22 July 2008 – estimated cost for the line was then £15.9 billion.

Designed to improve journey times across the capital, ease congestion and offer better connections, Crossrail will increase rail transport capacity by 10 per cent and bring an extra 1.5 million people to within forty-five minutes of central London. Construction work on Europe's largest civil-engineering project started in May 2009. With completion due in 2019 the total cost is expected to be £14.8 billion, more than £1 billion less than estimated in 2008.

The total length of the Crossrail project is eighty-five miles, extending from Reading and Heathrow Airport in the west to Canary Wharf and Abbey Wood in the southeast, and Shenfield in the northeast. Forty new stations are being built, ten of them underground, while the branch to Abbey Wood involves reusing part of the old North London Line through the Connaught Tunnel. Trains will be operated by sixty-five new 9-car Class '345' electric multiple units built by Bombardier at its Derby Litchurch Lane factory. Capable of 90 mph these trains each have a capacity of 1,500 passengers and will collect

their 25kV 50Hz AC current from overhead wires. Due to open in December 2018, the central section of Crossrail between Paddington and Whitechapel will see a service of twenty-four trains per hour, although opening of the entire route is not due to start until December 2019.

During the peak of construction there will be over 10,000 people working across forty sites while a total of eight enormous tunnel-boring machines have constructed twenty-six miles of new tunnels beneath central London. With names like *Phyllis*, *Ada*, *Sophia* and *Mary*, and operating in pairs driving twin parallel tunnels, these 500-ft-long, 23-ft-diameter German-built machines each weighed 1,000 tonnes and inched forward at the rate of 328-ft a week – as they moved forward, precast concrete segments were built in rings behind to form the tunnel lining.

Clearances were so tight in central London that in one place, at Tottenham Court Road station, one of the tunnelling machines was less than three feet above an existing Northern Line tunnel and three feet below an escalator. The excavated material from the tunnels was then shipped to Wallasea Island in Essex to create a new 1,500-acre RSPB nature reserve. A team of archaeologists have also been employed during the construction of Crossrail in central London and their finds include 5,000 human remains uncovered in the Bedlam burial ground beneath Liverpool Street station, many of them victims of the plague in the seventeenth century.

LEFT: Archaeologists at work on the Crossrail project, uncovering human remains in the Bedlam burial ground beneath Liverpool Street Station. Around 5,000 human remains, many dating from the Great Plague of the seventeenth century, have been discovered there.

BELOW: Watched by an excited crowd of construction workers, the 23-ft-diameter German-built tunnelling machine *Elizabeth* breaks through into Liverpool Street station in January 2015.

Modern motive power

It is becoming increasingly rare to see locomotive-hauled passenger trains on Britain's railways. With most passenger services now operated by diesel or electric multiple units, the exceptions are the sleeper trains between London and Cornwall and Scotland, an Arriva Trains service between Cardiff and Holyhead, Chiltern Railways between London Marylebone and Birmingham Snow Hill, and Virgin Trains East Coast between London King's Cross and Edinburgh. Last but not least are the Class '43' HST 125 power cars, introduced in 1975, which are still in active service with First Great Western, East Coast, East Midlands Trains, Grand Central and CrossCountry – the prototype of this class still holds the world speed record for diesel locomotives of 143 mph, set in 1973. Without doubt a British Rail success story

However, Britain's growing rail-freight industry employs a large number of mainly diesel-electric locomotives including, surprisingly, some classes that date back to the 1950s. Of the latter the humble Class '08' diesel shunters, introduced in 1952, were, until 2014, the most numerous with around 100 still operational around Britain. The oldest mainline diesel-electric locomotives still in operation are the English Electric Bo-Bo Class '20s' introduced in 1957, of which twenty survive and can still be seen working nuclear flask trains in pairs and the Rail Head Treatment Train during the winter. A number of Class '31' A1A-A1A, built by Brush Traction and first introduced in 1957, are also still in revenue-earning service, as is a fleet of refurbished Class '37' Co-Co diesels that were first introduced in 1960.

Other long-lived classes of diesel-electrics still on active service are Class '47' Co-Co locos built by Brush and BR Crewe Works and introduced in 1962, while a further thirty-three that were rebuilt as Class '57' are also still operational. A small number of the Class '56' Co-Co locomotives introduced between 1976 and 1984 still see service and more are earmarked for refurbishment.

The oldest electric locomotives still in operation are thirty-nine Class '73' Bo-Bo electro-diesels, which were first introduced in 1962 and are capable of working on both electrified third-rail and non-electrified routes. Some of these have recently been given a new lease of life for working sleeper trains north of Glasgow. Thirty of the Class '86' AC electric locos built in the mid-1960s still see service on the West Coast Main Line hauling Freightliner trains while several examples have been exported to Bulgaria and Hungary.

More modern heavy-freight diesel-electric locomotive classes delivered before privatization include fifteen of the North American EMD Class '59' Co-Co introduced in 1985 and 100 of the Brush-built Class '60' Co-Co introduced in 1989, although the latter class's numbers have dwindled in recent years.

Those delivered since privatization include 446 North American EMD Class '66' Co-Co, introduced in 1998 (the most numerous class on Britain's railways), thirty of the Spanish-built Class '67' Bo-Bo, introduced in 1999 and normally used on passenger services, twenty-five of the Spanish-built Class '68' Bo-Bo, introduced in 2014, and thirty of the US-built Class '70' Co-Co.

LEFT: Following a single round-trip trial run with coaching stock between Carlisle and Crewe, the two newly delivered DRS Vossloh 'Euro Light' Class '68s' destined for hire to Chiltern Railways head back light through Carlisle en route to Crewe Gresty Bridge depot on 21 August 2014.

ABOVE: During the short period of time when the Class '70s' could be seen operating into Scotland, Freightliner's No. 70010 heads a Ravenstruther to Drax power station loaded coal train at Mossband, near Gretna, on 8 April 2011.

Modern AC electric locomotives introduced before privatization are fifty of the Class '90' Bo-Bo built at BREL Crewe Works between 1987 and 1990, thirty-one of the Class '91' Bo-Bo, also built at BREL Crewe Works and introduced in 1988 for express passenger duties on the East Coast Main Line, and the forty-six Class '92' dual-voltage Co-Co, built jointly by Brush Traction and Asea Brown Boveri of Switzerland, which were introduced in 1993 specifically for hauling freight through the Channel Tunnel. The only new electric locomotives ordered since privatization are ten of the Spanish-built Class '88' electro-diesels that have yet to be delivered.

ROSCOs

Since privatization of Britain's railways in 1994 the locomotives and rolling stock operated by the Train Operating Companies (TOCs) and Freight Operating Companies (FOCs) have been leased from Rolling Stock Operating Companies (ROSCO). Today there are three ROSCOs: Porterbrook, Eversholt Rail Group and Angel Trains, all a very far cry from sixty years ago when British Railways (and the British taxpayer) owned its entire fleet of 19,000 steam locomotives.

It is also a sad fact that locomotive building in Britain – once suppliers of steam locomotive to much of the world – is now dead. Introduced in 1989 the Class 60 diesel-electrics built by Brush Traction were the last of their type to be built in Britain as were the Class 91 electric locomotives built by BREL Crewe Works and introduced in 1988. Since then all new mainline locomotives for Britain's railways have been built in North America or mainland Europe.

Modern freight services

One of the big post-privatization success stories on Britain's railways has been the growth of long-haul intermodal and trainload operations. Both recommended as the way forward in the 1963 'Beeching Report', intermodal 'liner trains' and merry-go-round coal trains today form a large part of railborne freight traffic. Intermodal trains carry internationally standardized road/rail/ship containers between the five deep-sea ports of Felixstowe, Seaforth (Liverpool), Southampton, Thamesport and Tilbury to inland terminals such as Daventry International Freight Terminal in the West Midlands, Trafford Park Euroterminal in Manchester and Coatbridge in Scotland – railborne intermodal traffic to and from the Continent via the Channel Tunnel is dealt with at Dollands Moor in Kent.

Trainload coal for coal-fired power stations still amounts to nearly 30 per cent of trainload freight although nearly all is now imported via ports such as Immingham in Lincolnshire and Hunterston in Scotland. Coal is also transported from Britain's last deep-mine colliery at Hatfield in South Yorkshire and opencast sites in South Wales and Scotland. Trainloads of imported biomass in the form of wood pellets is also increasingly being used by power stations instead of coal, leading to traffic flows from ports such as Tyne, Immingham and Portbury.

Supermarkets such as Tesco, Asda, Morrisons and M&S are increasingly using intermodal rail transport to shift groceries and drinks from the Daventry hub to depots in Scotland and Wales. Companies such as Ford, Honda, Jaguar, Landrover and BMW also use rail transport to convey new vehicles to ports such as Portbury for onward export by sea. The growing list of trainload freight operations also includes cement products from Lafarge plants, steel from the Tata plants at Margam to Llanwern in South Wales and from Scunthorpe to France via the Channel Tunnel, and last but not least 'binliner' trains of household waste from London and Bristol to landfill sites such as Stewartby in Bedfordshire and Calvert in Buckinghamshire. Intermodal traffic has doubled since privatization of the railways in 1994 and the amount of trainload freight carried by rail is also expected to continue growing for the foreseeable future.

Today there are five major freight-operating companies (FOCs) and several smaller concerns, all with their own leased fleets of powerful diesel and electric locomotives, moving intermodal containers, supermarket goods, coal, biomass, aggregates, china clay, minerals, timber, steel, scrap metal, new motor vehicles, oil, cement, household waste, railway infrastructure materials and nuclear waste around the country.

The largest of the FOCs is German-owned DB Schenker (DBS), the successor to the monolithic freight-operating company English Welsh & Scottish Railway which was formed following the privatization of BR. DBS currently employs more than 3,400 people in the UK, operating over 5,000 trains each month including services to and from mainland Europe via the Channel Tunnel and HS1 – moving intermodal containers, metals, coal, biomass and aggregates makes up much of their business. The company also supplies locomotives for the Royal Train, passenger charter work and hauled passenger services operated by First ScotRail, Chiltern Railways and Arriva Trains Wales.

Created by British Nuclear Fuels in 1995, Direct Rail Services is the only publicly owned rail FOC of any size in the UK. In addition to operating all the nuclear flask trains between nuclear power stations and the Sellafield reprocessing plant in Cumbria, and low-level radioactive waste to the Drigg repository, the company also operates intermodal trains on behalf of road operators such as Eddie Stobart. Dedicated intermodal trains for supermarket giants such as Tesco and Asda are now a regular sight on our railways.

GB Railfreight is another large operator of railfreight services in Britain having been purchased by Europorte, a subsidiary of French-owned Eurotunnel, in 2010. In addition to extensive intermodal, coal, aggregate and infrastructure traffic the company was awarded the contract to operate the Caledonian Sleeper service between London and Scotland in 2015.

The privatized successor to British Rail's Freightliner sector is the Freightliner Group – today a subsidiary of US company Genesee & Wyoming – which is the largest intermodal freight transport operator in Britain. In addition to a fleet of twenty-six former British Rail class 86 and 90 electric locomotives, the company also operates a large fleet of powerful modern diesel locomotives including 132 Class '66' and nineteen Class '70'.

French-owned Colas Rail not only operates timber trains from various locations in northern and southwest England to a large wood processing plant at Chirk on the English/Welsh border but also operates a large fleet of rail-infrastructure support locomotives and specialist equipment for tracklaying and maintenance work across Network Rail.

Mendip Rail is one of the smaller independent FOCs, currently operating stone trains from its quarries in Somerset with a fleet of eight Class '59' diesels. In 1986 it became the first private company to operate its own fleet of diesel locomotives on Britain's main rail network. A subsidiary of British American Railway Services, Devon & Cornwall Railways is the smallest of the railfreight operators in Britain, operating a small fleet of Class '31' and Class '56' diesel-electric locomotives.

LEFT: Hauled by Class '70' No. 70013, the 6Z68 Killoch to Cottam Freightliner Heavyhaul train of loaded coal hoppers, which would take the Settle–Carlisle route, crosses a remarkably calm and tidal River Esk at Mossband near Gretna on 6 February 2012.

ABOVE: Regular diagrammed loco for the service at the time and bearing the Stobart corporate livery, DB Schenker's No. 92017 *Bart the Engine* climbs the grade at Scout Green heading the 4S43 Rugby-Mossend Tesco intermodal on 31 July 2010.

Modern light railways

By the beginning of the twentieth century there were just over 300 public street tramways operating in most towns and cities throughout Britain. A few were steam powered, horse powered or even cable hauled but the vast majority were powered by electricity collected through an overhead wire. However, by the 1930s many of these localized tram systems were being replaced by more flexible trolley buses or motor buses. In London the last trams ran on 5 July 1952, while the last services ran elsewhere in Sunderland on 1 October 1954, Liverpool on 14 September 1957, Leeds on 7 November 1959, Sheffield in October 1960, Grimsby and Immingham on 1 July 1961 and, finally, Glasgow on 31 August 1962. After that date there was just one street tramway system operating in the entire country – along the seafront in Blackpool where it still operates today.

Today there are nine standard-gauge light-rail systems operating in England and Scotland with a total of 200¾ route miles. All are powered by electricity but not all are pure street tramways with some, such as the Manchester Metrolink, using converted railway routes as part of their network. Since the opening of the Tyne & Wear Metro in 1980 these modern light-rail systems have been a great success story and across the eight English systems there were 227 million passenger journeys taken in 2013/14. Almost 60 per cent of these journeys were on the two London networks. Networks continue to expand, particularly in Manchester in recent years, all helping to reduce congestion and vehicle pollution in city centres.

Docklands Light Railway
Length: 21 miles *Opening date:* 1987
Power supply: Third rail 750V DC
By 1980 London's redundant Docklands to the east of the City had become derelict. Over the following eighteen years an area of 8½ square miles was regenerated by the Government-supported London Docklands Development Corporation. The Docklands Light Railway is a completely automated light-metro system built to serve the area and was opened in three stages between 1987 and 2011. Serving the major financial centre of Canary Wharf much of it is elevated, with major sections built on redundant railway formations. The current network stretches from Bank station in the west to Beckton and Woolwich Arsenal in the east and from Stratford International in the north to Lewisham in the south. There are plans to extend the DLR to Dagenham Dock, Victoria, Charing Cross, Euston, St Pancras, Catford and Bromley North in the future.

Tramlink
Length: 17 miles *Opening date:* 2000
Power supply: Overhead 750V DC
Located in South London, Tramlink serves the boroughs of Croydon and Merton. Trams run on a mixture of track, on-streets, new alignments and former railway lines between Beckenham Junction, Elmers End, New Addington and Wimbledon. A number of extensions and new routes are planned for the future.

Nottingham Express Transit
Length: 9 miles *Opening date:* 2004
Power supply: Overhead 750V DC
Currently trams operate only between Hucknall, Phoenix Park and Station Street in Nottingham. Phase two between Station Street and Toton (six miles) and Station Street to Clifton (4½ miles) is currently under construction with part of the route using a former Great Central Railway formation.

Midland Metro
Length: 13 miles *Opening date:* 1999
Power supply: Overhead 750V DC
The majority of the tram route lies along the formation of the GWR mainline between Birmingham Snow Hill and Wolverhampton. An extension from Snow Hill to Edgbaston via Birmingham city centre is due to open in stages between 2015 and 2017. A further extension to Birmingham International Airport and Coventry is a serious contender for 2021.

LEFT: Set in a canyon of glass and steel, Canary Wharf station on the Docklands Light Railway serves one of the world's most important financial centres.

ABOVE: The Nottingham Express Transit is currently being expanded along two routes, one of which will connect with the planned East Midlands Hub station at Toton on the HS2 route to Leeds.

Sheffield Supertram
Length: 18 miles *Opening date:* 1994
Power supply: Overhead 750V DC
Serving the city of Sheffield and its environs, this tram system operates along a mixture of on-street, reserved rights of way and former railway alignments. It has three routes all running via the city centre: Yellow route Middlewood to Meadowhall Interchange; Blue Route Malin Bridge to Halfway; and Purple Route Meadowhall Interchange to Herdings Park. A new route between Meadowhall and Dore is planned for the future.

Tyne & Wear Metro

Length: 46 miles *Opening date:* 1980

Power supply: Overhead 1500V DC

Britain's first modern light-rail project, the initial section of the Tyne & Wear Metro was opened along the formation of the North Eastern Railway's Tyneside network that was originally electrified in 1904 – passenger services on parts of today's system first started as early as 1839. Diesel multiple units operated services from 1967 until the route was rebuilt as a light railway and re-electrified. Opened in stages between 1980 and 2008, the metro has two distinct routes: Yellow Line between St James and South Shields includes the North Tyneside Loop via Whitley Bay; and Green Line between Newcastle Airport and South Hylton.

Manchester Metrolink

Length: 57 miles *Opening date:* 1992

Power supply: Overhead 750V DC

The most extensive of Britain's light-rail systems, the Manchester Metrolink now boasts seven routes serving ninety-two stations that were opened in stages between 1992 and 2014. Trams use a mixture of on-street tracks, new formations and former railway lines to serve Bury and Rochdale in the north, Ashton-under-Lyne in the east and Eccles, Altrincham, East Didsbury and Manchester Airport in the south. A planned new 3½-mile route between Port Salford and Pomona is due for completion in 2019.

Blackpool Tramway

Length: 11 miles *Opening date:* 1885

Power supply: Overhead 600V DC

A national institution, the 130-year-old Blackpool Tramway is the only street tramway in Britain never to have closed and one of only three in the world to use double-deck tramcars. Trams mainly run on a dedicated roadside formation along the seafront between Starr Gate in the south and Fleetwood Ferry in the north. The operational fleet consists of modern Flexity 2 single-deck tramcars and nine refurbished double-deck Balloon Cars dating from 1934. A heritage fleet of older tramcars can be seen at work during the peak summer months and during the famous illuminations.

Edinburgh Trams

Length: 8¾ miles *Opening date:* 2014

Power supply: Overhead 750V DC

Last but not least is the newest light-rail system to open in Britain. Trams operate on a mixture of new formations and on-street sections between Edinburgh Airport and York Place in New Town. A sorry story of mismanagement, contractual disputes, and cost and completion overruns has meant that the citizens of Edinburgh never got the entire route that was planned – despite preliminary roadworks that gridlocked the city centre, the section from York Place to Newhaven via Leith was never built. Costs have risen from the original estimate of £375 million to a likely bill, including interest, of at least £1 billion.

LEFT: At fifty-seven miles in length the Manchester Metrolink network is Britain's most extensive light-rail system, much of it built along disused former railway routes around the city.

BELOW: After years of disruption the city of Edinburgh finally got its trams in 2014, but at great cost to the taxpayer and with a route that is much shorter than originally planned.

Mainline steam tours

The end of standard-gauge steam haulage on British Rail at midnight on 11 August 1968 was a poignant moment for railway enthusiasts across Britain. Despite a complete ban on steam after that date one locomotive had managed to defy it – ex-LNER Class 'A3' 4-6-2 *Flying Scotsman* had been saved from the scrapheap by businessman Alan Pegler in 1963, who then lavished large amounts of money on having it restored at Doncaster Works. Pegler had been running railway enthusiasts' specials since the early 1950s and his connections with the top brass at British Railways gave him the opportunity to operate *Flying Scotsman*-hauled specials on the mainline both before and after the ban on steam. Although this agreement with BR ran until 1972, these steam operations ended when the locomotive was shipped off to the USA in 1969 for an eventful tour to promote British exports – before this the last enthusiasts' tour headed by *Flying Scotsman* was a return trip between London King's Cross and Newcastle on 31 August 1969. More than two years elapsed before steam was to return to the mainline.

Meanwhile, one more famous steam locomotive was waiting in the wings to make its return to mainline running. Withdrawn in 1962, ex-GWR 4-6-0 *King George V* had been preserved as part of the National Collection and was being restored to working order at the Bulmer's Railway Centre in Hereford. On October 1971 it made a triumphant return to mainline running, hauling a series of enthusiasts' specials. This very first steam-hauled tour since the ban on steam haulage took in a route from Hereford to Tyseley via Severn Tunnel Junction, Swindon and Oxford. Three other tours followed during that week including a visit to Kensington Olympia in London.

Since those early days of the return of steam over forty years ago, there has been a massive increase in the operation of steam-hauled tours around Britain. Locomotives that were once rusting hulks have been lovingly restored to working order and rarely a week goes by without a steam train operating somewhere on Britain's rail network. In addition to these tours a regular timetabled steam train has been operating from May to

October since 1984 on the Mallaig Extension of the West Highland Line in Scotland – 'The Jacobite' has become so popular that a second service operating from June to August was added in 2011.

Britain is fortunate to possess a plethora of scenic rail routes, some of which tax the skills of steam-locomotive drivers and firemen to the limit. Notable among these is the steeply graded 73-mile Settle–Carlisle Line in northern England which now sees steam-hauled trains on a regular basis all year round – 'The Dalesman' and 'The Fellsman' trains are weekly performers during the summer months. The scenic delights of the former-GWR coastal route between Exeter and Paignton and the Central Wales Line between Llanelli and Craven Arms also see regular steam trains during the summer months. Other popular routes include York to Scarborough, London to the Dorset Coast and Birmingham to Stratford-upon-Avon while during the pre-Christmas period there are steam tours to Christmas Markets at various destinations such as Bristol, Bath, Chester and Lincoln.

Operators of steam tours now have a large pool of powerful preserved steam locomotives to draw on, with examples ranging from GWR 'Castle' and 'King' Class 4-6-0s, LMS 'Coronation' Class and 'Princess Royal' Class 4-6-2s, LNER 'A4' Class 4-6-2, Southern Railway Bulleid Pacifics and BR Standard Class 4-6-2s. A newcomer in 2009 was the newly built Peppercorn 'A1' Class 4-6-2 *Tornado*, which made its mainline debut on 31 January

LEFT: BR Standard Class '8P' 4-6-2 No. 71000 *Duke of Gloucester* in charge of the northbound 'Cumbrian Mountain Express' to Carlisle at Selside on the Settle–Carlisle Line on 19 May 2011.

BELOW: A classic autumnal afternoon was enjoyed by the passengers of the 'Scottish Lowlander' railtour on 27 September 2014 as it journeyed the equally classic Glasgow & South Western route back via Dumfries to Carlisle, where 'Coronation' class Pacific 46233 *Duchess of Sutherland* took over to Crewe. 'Gresley' Pacific 60009 *Union of South Africa* is seen on the final assault of the 1 in 200 climb up to Enterkinfoot tunnel. Until 1923 the line via Dumfries was in competition with the North British Railway and Caledonian Railway as one of the main lines into Scotland.

with a return trip between York and Newcastle. Following a recent overhaul this fine locomotive will be a regular performer for many years.

Although steam locomotives certified for mainline running are limited to a maximum speed of 75 mph, an exception was made in 2013 when ex-LNER 'A4' *Bittern* ran special charter trains on the East Coast Main Line to celebrate the seventy-fifth anniversary of sister engine *Mallard* making its 126 mph world-record-breaking run in 1938 – on one of its runs 76-year-old *Bittern* achieved a top speed of 93 mph, a new speed record for a British preserved steam locomotive.

During 2015 there were around 300 steam-hauled tours scheduled to operate around Britain, and barring any unforeseen problems this level is likely to continue for some years. There are currently twenty-five tour operators ranging from Belmond, who organize luxury Pullman dining trains, to the Railway Touring Company who organize, amongst others, the 9-day 'The Great Britain' which is now in its eighth year of operation. By far the most extensive steam tour, 'The Great Britain' is hauled over each stage by an array of different locomotives – the 2015 tour ranged far and wide, starting and ending in London, and including journeys to Par in Cornwall, Bristol, Hereford, Grange-over-Sands in Cumbria, the Cumbrian Coast route, Edinburgh, Aberdeen, Inverness, Kyle of Lochalsh, Wick, Aviemore, Manchester and Grantham. How the railway scene has changed since those dark days in August 1968!

Sadly, a serious incident on 7 March 2015 put steam charter train operations in turmoil. Just over two months short of the centenary of the tragedy on 22 May 1915 when 225 people were killed in a multiple collision at Quintinshill near Gretna Green, a steam charter train headed by 'Battle of Britain' Class 4-6-2 No. 34067 *Tangmere* passed a red signal and ended up straddled across a busy junction at Wootton Bassett, west of Swindon. An HST 125 had passed through the junction at high speed less than one minute before but fortunately no other trains were involved. The operator of the charter train, West Coast Railways, was then immediately ordered to suspend operations by Network Rail until a number of steps were taken to satisfy safety regulations. It is fairly obvious that any serious accident involving a steam charter train would bring these operations to an abrupt and permanent end.

The rising sun has barely shown itself over the Ben Alder mountain range and the clouds are already starting to close in as Stanier 'Black 5' 4-6-0 44871 climbs up towards Corrour station beneath the crystal clear morning sky across Rannoch Moor on 16 October 2010.

44871

Railway footpaths and cycleways

Since nationalization in 1948 over 7,000 route miles of railways throughout Britain have been closed. Around 3,000 miles of loss-making rural lines had been closed by the end of 1962 before the second wave of closures of around 4,000 miles that followed the publication of the 'Beeching Report' in 1963. By the early 1970s vast swathes of Britain were without a rail link to the outside world for the first time in over 100 years. However, by the 1980s the closures were at an end, thanks to politicians' more enlightened views on public transport. There has even been a reversal in fortunes over the last forty years with around 215 miles of track and 360 stations reopened but, although good news, this is just a drop in the ocean compared to the thousands of miles of lost railways that criss-cross Britain.

As we have read in the previous chapter, a good number of previously closed lines have been reopened as heritage railways and many of those today are thriving concerns attracting huge numbers of visitors each year, in turn supporting local tourist economies. However, the total length of these railways, both standard- and narrow-gauge is just over 500 route miles.

Despite rail reopenings of both kinds there still remains probably around 6,300 miles of lost railways across the country. What has happened to them? A close examination of 1:50,000 Ordnance Survey maps will usually give clues such as embankments, cuttings, viaducts and bridges that trace the route of a long-closed railway across the landscape. On the ground, nature has usually taken over and long rows of trees often are a giveaway to the route of a long-lost railway. Many of the substantial station buildings have found a new lease of life as private residences complete with platforms that are usually screened from public gaze, while substantial stone viaducts and iron bridges are a fitting memorial to the nineteenth-century railway builders.

Fortunately not all of Britain's lost railways have disappeared. The process of breathing new life into them started nearly eighty years ago when the LMS gave the track bed of the closed narrow-gauge Leek & Manifold Valley Light Railway in 1937 to Staffordshire County Council for use as a footpath and bridleway. In more recent times, while many old railway routes have been lost to road improvements and urban or industrial development, there has been a more enlightened and refreshing approach to the use of these wildlife corridors. Many have been reopened as footpaths by local councils while the sustainable transport charity Sustrans has gone much further by resurfacing a large number of them as traffic-free routes for cyclists. The first of these was the Bristol to Bath Railway Path, which was reopened between 1979 and 1986, and since then the charity has opened over 10,000 miles of mainly traffic-free routes, many of them along the trackbeds of closed railways. The National Cycle Network sees well in excess of 400 million journeys taken by cyclists each year and this number is likely to rise as more routes are opened. Sustrans, along with local landowners and councils, have provided signposts, stylish mileposts, information boards and car parks. Some lost railways have also been incorporated into long-distance footpaths (LDP) – much of the old railway route between Nethy Bridge and Craigellachie in northeast Scotland today forms part of the Speyside Way LDP. In addition to human recreational activities these lost railways are also important wildlife corridors where birds, butterflies, insects, small mammals and wild flowers all flourish.

While most of these footpaths and cycleways can be found in the countryside or along the coast, a few have also found favour with city-dwellers and cycling commuters. Notable among these are the Deeside Way on

the western outskirts of Aberdeen, the Bristol and Bath Railway Path, the Parkland Walk between Finsbury Park, Highgate and Alexandra Palace in North London and the Innocent Railway Path in Edinburgh – the latter skirts the southern slopes of Holyrood Park and features Scotland's oldest railway tunnel at St Leonards, which was built in 1830.

Wherever you may be in Britain there is a railway path and cycleway not far away. They all have their own unique character, with many incorporating dark railway tunnels, viaducts and bridges – details of fifty of these can be found in the author's *Exploring Britain's Lost Railways*, published by HarperCollins.

LEFT: A 4-mile section of the Cockermouth, Keswick & Penrith Railway trackbed between Keswick and Threlkeld is now a popular traffic-free footpath and cycleway known as the Keswick Railway Path.

BELOW: Walking in the wilds of Dartmoor – this remote granite bridge spans the trackbed of the former Yelverton to Princetown branch line which closed on 3 March 1956.

Railway reopenings

As we have read, railway closures continued unabated between 1948 and the mid-1970s, and during that dark period it would have been difficult, nigh impossible, to have been optimistic about the future of Britain's railways. However, once the dust had settled, concerns for the environment and about road congestion slowly led to a more enlightened attitude to public transport, in particular the future role of railways. Even before privatization in 1994, railway and station reopenings had started to become a reality, although the vast majority of these were on routes that still supported rail freight operations.

One of the first of these was the reopening to passengers of the line between Peterborough and Spalding in 1971, while in Scotland one of the first reopenings came about following the closure of the Edinburgh to Perth mainline to make way for the new M90 motorway in 1970 – the section closed was between Cowdenbeath and Bridge of Earn via Kinross meaning that trains between Edinburgh and Perth were forced to take a much longer route via Larbert and Stirling. To shorten this journey the freight-only single-track route between Ladybank and Hilton Junction was reopened to passenger services in 1975.

The 1980s saw a rash of railway reopenings including Barnsley to Penistone (1983), Edinburgh to Bathgate (1986), Oxford to Bicester (1987), Morecambe to Heysham (1987), Coventry to Nuneaton and Coventry to Leamington Spa (both 1987), Kettering to Corby (1987, closed again 1990, reopened 2009), Abercynon to Aberdare (1988), Walsall to Hednesford (1989, extended to Rugeley in 1997) and Shields Junction to Paisley Canal (1990). In South Wales the Maesteg Line north from Bridgend reopened in 1992. The only brand-new railways to open during this late-BR period were the 13¾-mile Selby cut-off of the East Coast Main Line in 1983 and the short links to Stansted Airport in 1991 and Manchester Airport in 1993.

Railway reopenings gathered pace, particularly in Wales and Scotland, following privatization in 1994. In England the Blackburn to Clitheroe line reopened in that year while the Robin Hood Line between Worksop and Nottingham was completed in 1998. In Wales the Vale of Glamorgan Line between Barry and Bridgend reopened in 2005 and the Ebbw Vale Line between Ebbw Junction, south of Newport, and Ebbw Vale in 2008. In London the 53½-mile London Overground network was created between 2007 and 2014 by linking existing passenger

routes by way of previously freight-only lines such as between Kensington Olympia and Willesden Junction.

The only brand-new railways built since privatization have been the Heathrow Airport Link from Paddington, which opened in 1998, and the 68-mile High Speed 1 (HS1) Channel Tunnel rail link between Folkestone International and London St Pancras, which was completed in 2007.

In Scotland the Stirling to Alloa line opened in 2008 and the electrified Airdrie to Bathgate Link opened in 2010, giving a third rail route between Edinburgh and Glasgow. The biggest success story of railway reopenings in Britain must be the Borders Railway, which reopened on 6 September 2015. The 30-mile line was rebuilt along the northern section of the Waverley Route that closed between Edinburgh and Carlisle in January 1969, leaving the Borders region bereft of rail connections to the outside world. Linking pre-existing stations at Edinburgh Waverley, Brunstane and Newcraighall, the railway serves new stations at Shawfair, Eskbank, Newtongrange, Gorebridge, Stow, Galashiels and Tweedbank.

And what of the future for railway reopenings? For many years rail users on the Far North of Scotland route between Inverness and Wick/Thurso have been lobbying for the Dornoch Rail Link that would considerably shorten journey times between these two destinations. The link would cross the Dornoch Firth between Tain and Golspie via Dornoch but it would mean closing the 46-mile meandering inland route via the town of Lairg.

Down in South Devon the damage to the coastal railway at Dawlish during the storms of February 2014 has once again sparked the demand for an alternative route between Exeter and Plymouth. Of course one used to exist but it was closed as a through route in 1968 and now only Bere Alston (for the Gunnislake branch) in the south and Meldon (Okehampton) in the north are still rail served. There are already serious proposals to extend the line north from Bere Alston to Tavistock by Devon County Council, and extending this around Dartmoor to Meldon would give an alternative rail link between Exeter and Plymouth. Watch this space!

LEFT: Borders Railway progress – after starting tracklaying operations at Millerhill in the morning, the tracklaying train had reached Shawfair by the end of the day, 6 October 2014. GBRf Class '66', No. 66736 *Wolverhampton Wanderers*, propels the train into Shawfair station over the previously laid section.

ABOVE: On 15 April 2015 Future Rail contractors are undertaking final track fixing and gauge verification work at the Tweed river bridge, on the section of the reinstated Waverley Line between Galashiels and Tweedbank stations. A new illuminated cycleway/footpath has also been installed alongside the railway, providing a safe and direct route to and from Galashiels for the residents of Tweedbank.

The future

What does the future hold for Britain's railways? As we have already read, the £14.8 billion Crossrail project in London will be completed by 2019, electrification of the Great Western main line out of Paddington by 2017, the Midland main line out of St Pancras and the Northern Hub both by 2020. This is all welcome news but where Britain lags miserably behind the rest of the world is in the lack of investment in dedicated high-speed railways.

To put this into context Britain, the birthplace of railways, currently has only one dedicated high-speed route – the 67-mile HS1 between Folkestone and London St Pancras, which was completed in 2007. Elsewhere in the world the picture is very different:

- Japan has 1,968 miles of Shinkansen lines either already opened, in the course of construction or planned. The country led the world with the opening of the 320-mile Tokaido Shinkansen line in 1964
- France has 1,718 miles of LGV routes either already opened or being constructed. The first dedicated high-speed route in Europe was the 254-mile LGV Sud-Est which opened in 1981
- Germany has 643 miles, of which the first was the 160-mile Berlin to Hanover route that opened in 1998
- Spain already has 1,407 miles of dedicated high-speed routes while Italy has 545 miles and Belgium 154 miles
- Turkey will have 521 miles of dedicated high-speed railways by 2016
- China already has 9,900 miles of high-speed routes, with at least the same mileage being planned for the future

So why does Britain lag so far behind and what is being done about it?

Britain's existing rail network was built during the Victorian era, with many routes and their infrastructure dating back well over 150 years. While upgrading the existing main lines for faster trains has been an ongoing process since the 1960s, there is a limit to what can be done without massive disruption to services. The West Coast Main Line, Britain's premier railway route, is a classic example – first upgraded during electrification between the late 1950s and 1974 and then by Railtrack/Network Rail in the early years of the twenty-first century, it has probably reached its limit both for speed and line capacity. Railtrack's ambitious plan to increase line speed to 140 mph never materialized and the project was dogged by mismanagement, a cost overrun of more than 400 per cent and completion three years late.

With passenger numbers and freight operations forecast to continue growing in the foreseeable future, the only practical way forward must be to build a network of dedicated high-speed routes connecting Britain's main areas of population. This will then theoretically free up capacity on the existing network for more freight

operations. A start, of sorts, has already been made with Phase 1 of the High Speed 2 (HS2) line between London, Birmingham and a connection with the West Coast Main Line near Rugeley in Staffordshire. In worldwide terms it is but a drop in the ocean but the 119-mile electrified route would see trains operating at up to 250 mph and cut journey times between the capital and Birmingham to forty-nine minutes, twenty-three minutes less than today's fastest time. Phase 1 has already been approved and construction work is due to start in 2017 with completion in 2026. Of course one thing that has not changed in over 150 years is the reaction to the coming of this new railway to hitherto peaceful stretches of English countryside – as in the 1830s when the London & Birmingham Railway was being built, landowners and the middle classes of the Home Counties are still protesting vociferously about its projected route.

Phase 2 of HS2 extends Phase 1 from north of Lichfield to Manchester and Liverpool. Leeds will be served by a new line branching off Phase 1 to the east of Birmingham. On this line there will be new stations at East Midlands (Toton) and Sheffield (Meadowhall), while a spur will connect it to the East Coast Main Line south of York. Completion of Phase 2 is due by 2033.

The cost of HS2? Depending on whom you listen to it could be anything from £43 billion to £80 billion. Plans to extend it northwards to serve Glasgow and Edinburgh are presently just a pipe dream although the Scottish Government has already announced that they want to build a 46-mile dedicated high-speed rail link between the two cities by 2024, two years before the completion of Phase 1 of HS2.

LEFT: Map of the HS2 linking London Euston with Birmingham, Leeds and Manchester. Assuming that it is all completed by 2033, high-speed trains will be able to continue their journeys northwards via connections to the West Coast and East Coast Main Lines.

ABOVE: With a maximum speed of 250 mph, the new trains for HS2 could be built to a European loading gauge, making them wider and taller and thus able to carry more passengers. However, the existing lines that are proposed to be used by them – West Coast and East Coast Main Lines to Scotland – would need to be cleared to their loading gauge.

Index

Acknowledgements

t = top; b = bottom; r = right; l = left; m = middle

Photo credits:
Henry Casserley: 138; 160; 206; 207
Colour-Rail: 129 (Gordon Edgar); 146t; 162; 210 (D. Ovenden); 221ml; 225; 231; 234; 235; 236; 246/247 (T. Owen); 260 (R. Patterson); 261 (A. Gray); 264bl; 275 (Chris Milner); 278 (Bob Sweet); 288; 296 (Peter Delaney)
Ewan Crawford: 51b
Crossrail Ltd: 280; 281
Gordon Edgar: 8/9; 102; 123; 132; 161; 228; 233; 248; 254/255; 259; 262; 263b; 269; 270; 271; 272/273; 276; 282; 283; 284; 290; 291; 292/293; 297
Mike Esau/John Ashman: 243
John Goss: 151
Grace's Guide: 218bl
G.T. Hancock (courtesy Ashover Light Railway Society): 188/189
Tony Harden: 172; 186
Julian Holland: 205; 208bl; 208br; 209 bl; 209br; 230; 237; 242bl; 242br; 245; 252; 257; 295
HS2: 298; 299
ImageRail: 277
Steven Mackay: 279
Michael Mensing: 173
Milepost 92½: 294
Gavin Morrison: 190
Reproduced by permission of the National Library of Scotland: 99
Science & Society Picture Library: Front cover (NRM/Pictorial Collection); 7 (Colin T. Gifford); 10/11 (NRM/Pictorial Collection); 12 (Science Museum); 13 (National Railway Museum); 14/15 (Science Museum); 16 (NRM/Pictorial Collection); 17 (National Railway Museum); 18/19 (National Railway Museum); 20 (National Railway Museum); 21 (Science Museum); 22/23 (Science Museum); 24 (Science Museum); 25 (Science Museum); 26/27 (Science Museum); 28/29b (Science Museum); 29t (NRM/Pictorial Collection); 30 (National Railway Museum); 31 (NRM/Pictorial Collection); 32/33 (NRM/Pictorial Collection); 34/35b (NRM/Pictorial Collection); 34mr (National Railway Museum); 35mr (Science Museum); 36ml (National Railway Museum); 36b (NRM/Pictorial Collection); 37 (Science Museum); 38 (National Railway Museum); 39 (NRM/Pictorial Collection); 40 (NRM/Pictorial Collection); 41 (NRM/Pictorial Collection); 42 (NRM/Pictorial Collection); 43 (Science Museum); 44 (NRM/Pictorial Collection); 45tl (Science Museum); 45tr (NRM/Pictorial Collection); 46 (Science Museum); 47 (Science Museum); 48/49 (Science Museum) 51t (Science Museum); 52 (NRM/Pictorial Collection); 53 (National Railway Museum); 54/55 (NRM/Pictorial Collection); 56 (Science Museum); 57t (NRM/Pictorial Collection); 57b (NRM/Pictorial Collection); 58 (NRM/Pictorial Collection); 59 (NRM/Pictorial Collection); 60 (NRM/Pictorial Collection); 61 (NRM/Pictorial Collection); 62 (NRM/Pictorial Collection); 63 (Science Museum); 64 (NRM/Pictorial Collection); 65 (Science Museum); 66/67 (NRM/Pictorial Collection); 68 (Science Museum); 69 (Science Museum); 70 (Science Museum); 71 (Science Museum); 73 (National Railway Museum); 74 (Science Museum); 75 (National Railway Museum); 76 (Science Museum); 77 (Science Museum); 78 (National Railway Museum); 79 (Royal Photographic Society/National Media Museum); 80 (Science Museum); 82 (National Railway Museum); 83t (NRM/Pictorial Collection); 83b (National Railway Museum); 84 (Science Museum); 85 (Science Museum); 86 (NRM/Pictorial Collection); 87 (Science Museum); 88 (NRM/Pictorial Collection); 89 (National Railway Museum); 90bl (NRM/Pictorial Collection); 90br (Science Museum); 91 (NRM/Pictorial Collection); 92/93 (Science Museum); 94 (NRM/Pictorial Collection); 95 (NRM/Pictorial Collection); 96 (NRM/Pictorial Collection); 97 (NRM/Pictorial Collection); 98 (NRM/Pictorial Collection); 100 (NRM/Pictorial Collection); 101 (National Railway Museum); 103 (National Railway Museum); 104 (National Railway Museum); 105 (Colin T. Gifford); 106 (National Railway Museum); 107t (NRM/Pictorial Collection); 107ml (National Railway Museum); 108 (Past Pix); 109 (Science Museum); 110 (NRM/Pictorial Collection); 111 (National Railway Museum); 112 (National Railway Museum); 113 (NRM/Pictorial Collection); 114 (Science Museum); 115 (NRM/Pictorial Collection); 116 (Science Museum); 117 (National Railway Museum); 118 (National Railway Museum); 119 (National Railway Museum); 120 (National Railway Museum); 121 (National Railway Museum); 122 (NRM/Pictorial Collection); 124 (National Railway Museum); 125 (Past Pix); 127 (Science Museum); 128 (NRM/Pictorial Collection); 130 (National Railway Museum); 131 (NRM/Pictorial Collection); 134 (National Railway Museum); 135tl (NRM/Pictorial Collection); 135tr (NRM/Pictorial Collection); 136 (National Railway Museum); 137 (Colin T. Gifford); 139 (National Railway Museum); 140bl (NRM/Pictorial Collection); 140br (NRM/Pictorial Collection); 141 (NRM/Pictorial Collection); 143 (Daily Herald Archive/National Media Museum); 144 (National Railway Museum); 145 (National Railway Museum); 146b (National Railway Museum); 147 (National Railway Museum); 148/149 (National Railway Museum); 150 (NRM/Pictorial Collection); 152 (NRM); 153 (National Railway Museum); 154 (NRM/Pictorial Collection); 155 (NRM/Pictorial Collection); 156/157 (National Railway Museum); 159 (Daily Herald Archive/National Media Museum); 163 (NRM/Pictorial Collection); 164 (National Railway Museum); 165 (NRM/Pictorial Collection); 166 (NRM/Pictorial Collection); 167 (Colin T. Gifford); 168 (National Railway Museum; 169 (National Railway Museum); 171 (NRM/Pictorial Collection); 174 (National Railway Museum); 175 (National Railway Museum); 176t (Daily Herald Archive/National Media Museum); 176b (National Railway Museum); 178 (National Railway Museum); 179 (National Railway Museum); 180/181 (National Railway Museum); 182 (NRM); 183 (National Railway Museum); 184 (Daily Herald Archive/National Media Museum); 185t (NRM/Pictorial Collection); 185b (NRM/Pictorial Collection); 187 (NRM/Pictorial Collection); 191 (National Railway Museum); 193 (Manchester Daily Express/Science Museum); 194 (National Railway Museum); 195t (National Railway Museum); 195b (National Railway Museum); 196 (Daily Herald Archive/National Media Museum); 197 (NRM/Pictorial Collection); 198 (National Railway Museum); 199 (National Railway Museum); 200 (National Railway Museum); 201 (National Railway Museum; 202 (National Railway Museum); 203 (National Railway Museum); 204 (National Railway Museum); 211 (NRM/Pictorial Collection); 212 (National Railway Museum); 213 (NRM/Pictorial Collection); 214 (National Railway Museum); 215t (National Railway Museum); 215mr (NRM/Pictorial Collection); 216 (National Railway Museum); 217 (Planet News); 218br (Science Museum); 219 (National Railway Museum); 220bl (National Railway Museum); 220br (National Railway Museum); 221b (National Railway Museum); 222/223 (Daily Herald Archive/National Media Museum); 224bl (National Railway Museum); 224br (National Railway Museum); 226/227 (National Railway Museum); 229tl (National Railway Museum); 229tr (National Railway Museum; 232 (National Railway Museum); 238 (NRM/Pictorial Collection); 239 (Cuneo Fine Arts/National Railway Museum); 240bl (NRM/Pictorial Collection); 240br (NRM/Pictorial Collection); 241 (Cuneo Fine Arts/National Railway Museum); 244 (Colin T. Gifford); 249bl (National Railway Museum); 249br (NRM); 250 (NRM/Pictorial Collection); 251 (National Railway Museum); 253 (National Railway Museum); 256 (Daily Herald Archive/National Media Museum); 258 (National Railway Museum); 263mr (Manchester Daily Express); 264br (NRM/Pictorial Collection); 265 (National Railway Museum); 266 (NRM/Pictorial Collection); 267 (Cuneo Fine Arts/National Railway Museum); Back cover (Colin T. Gifford)
Shutterstock.com: 4/5 (Christophe Jossic); 286 (Kiev.Victor); 289 (Frank Gaertner)
Tramlink Nottingham: 287
Wikipedia: 81 (Chowells~commonswiki); 177; 192

With thanks for research assistance and advice to:
Gordon Edgar

Front cover:
Norman Wilkinson's iconic oil painting from 1937 of the streamlined 'The Coronation Scot' express ascending Shap Fell in Cumbria. It was used on a poster produced for the London, Midland & Scottish Railway to promote the new service between London (Euston) and Glasgow (Central).

Back cover:
Ex-Caledonian Railway '2P' 0-4-4 No. 55173 stands at Killin station with the branch train from Killin Junction in September 1961.